One True Logic

One True Logic

A Monist Manifesto

OWEN GRIFFITHS
and
A.C. PASEAU

OXFORD
UNIVERSITY PRESS

Great Clarendon Street, Oxford, OX2 6DP,
United Kingdom

Oxford University Press is a department of the University of Oxford.
It furthers the University's objective of excellence in research, scholarship,
and education by publishing worldwide. Oxford is a registered trade mark of
Oxford University Press in the UK and in certain other countries

First Edition published in 2022

Impression: 1

Published in the United States of America by Oxford University Press
198 Madison Avenue, New York, NY 10016, United States of America

British Library Cataloguing in Publication Data
Data available

Library of Congress Control Number: 2022931794

ISBN 978-0-19-882971-3

DOI: 10.1093/oso/9780198829713.001.0001

Printed and bound in Great Britain by
Clays Ltd, Elcograf S.p.A.

Acknowledgements

We are grateful to many readers and audiences over the years. AJ Gilbert, Alex Oliver, Gila Sher, Hans Robin Solberg, Michael Bevan, Rob Trueman, Tim Williamson, Weng Kin San, and Wes Wrigley read the entire manuscript in draft and provided excellent comments and suggestions. Michael also lent us invaluable research assistance, twice over, in the summers of 2018 and 2019, and compiled the first version of the index. §4.4 draws extensively on research he carried out for us. We benefited from our editor Peter Momtchiloff's good-humoured guidance throughout, as well as from the comments of Ole Hjortland and Simon Hewitt for Oxford University Press.

We tried out early versions of the material in this book on students in a BPhil class we jointly taught in 2017 and again in 2018. We are grateful to all these students for their feedback, in particular AJ Gilbert, Kevin Gibbons, Michael Bevan, Oliver Hurcum, Sabine Bot, Sean McIntosh, Tomi Francis, and Weng Kin San. Jack Woods and Paul Shafer later invited us to Leeds to give two talks in May 2019. We are grateful for the invitation and thank audience members at the Logic Seminar and the Centre for Metaphysics and Mind for comments, especially Robbie Williams and Jack.

§§1.4–1.6 and 3.4 include revised material from Owen's 2013 paper 'Problems for Logical Pluralism', published in *History and Philosophy of Logic*. He is grateful to Alex Oliver and Florian Steinberger for comments on drafts of that paper, as well as audiences in Cambridge and London.

Owen presented an early version of Chapter 3 at the Logic, Epistemology and Metaphysics Seminar at London's Institute of Philosophy in 2017, and a later version at the Serious Metaphysics Group in Cambridge in 2020. Thanks respectively to Florian Steinberger, and to Wouter Cohen and Alexander Bird for the invitations.

Alex presented Chapter 5 at the Oxford Philosophy of Mathematics seminar as well as the Kurt Gödel Research Centre in Vienna in 2018, and a French version that same year at the École d'Été in St Flour. Thanks respectively to Dan Isaacson and Volker Halbach, Neil Barton and Sy Friedman, and Brice Halimi for these invitations, and to the various audience members. §5.3 overlaps with our article 'Is English Consequence Compact?' published in *Thought*, whose referees we thank for their comments.

Chapter 9 is a revised version of our 2016 *Bulletin of Symbolic Logic* article 'Isomorphism Invariance and Overgeneration'. We would like to thank Patricia Blanchette, Denis Bonnay, Tim Button, Solomon Feferman, Salvatore Florio, Gila

Sher and an anonymous referee for their helpful comments on previous drafts of that paper. We received similarly helpful feedback from audiences in Cambridge, Glasgow, London, Munich, and Oxford. §§6.5 and 8.2 were improved following comments on the latter by Tim Button and Sean Walsh, and Denis Bonnay's comments improved §§8.3 and 8.4.

Alex would like to thank Wadham College and Oxford's Philosophy Faculty for their support throughout the conception and writing of this book. Owen thanks Churchill College and Cambridge's Faculty of Philosophy. We would like to thank Marta Weiss for sourcing the cover image.

Alex dedicates this book to his parents, Antoine and Marina. Owen dedicates it to his parents, Martyn and Roma. Since collective attributions are not reducible to individual ones, that leaves us with one last task. We jointly dedicate this book to our former supervisor, Alex Oliver.

Contents

PART II. THE L∞G∞S HYPOTHESIS

PART III. OBJECTIONS

Introduction

Logical consequence is the central concept of logic. The ideas of a sentence's follow-ing from some others and of an argument's being valid are appealed to constantly in philosophical, mathematical, scientific, legal, and everyday discourse. In 1936, Alfred Tarski published his paper 'On the concept of logical consequence', which introduced the now-standard model-theoretic definition of logical consequence. John Etchemendy summarizes the significance of this work:

> The highest compliment that can be paid to the author of a piece of conceptual analysis comes not when the suggested definition survives whatever criticism may be levelled against it, or when the analysis is claimed unassailable. The highest compliment comes when the suggested definition is no longer even seen as the result of conceptual analysis—when the need for analysis is forgotten, and the definition is treated as common knowledge. Tarski's account of the concepts of logical truth and logical consequence has earned him this compliment.
>
> (Etchemendy 1990, p. 1)

Etchemendy is correct about the status of Tarski's early model-theoretic defini-tions. There is some question about the correct interpretation of Tarski—especially whether his models all have the same domain—but his definition is more or less accepted by most. Certainly, his are the definitions that we typically teach to beginning logic students, and most never encounter reasons to deviate.

Etchemendy is also correct that Tarski's definitions are no longer viewed as the results of conceptual analysis. Yet Tarski was clear that, 'in defining this concept, efforts were made to adhere to the common usage of the language in everyday life' (1936, p. 409). It is striking that, for surely one of the most successful and widely accepted pieces of analysis, conceptual or otherwise, there is so little discussion of the everyday usage of logical consequence and the extent to which Tarski succeeded in capturing it.

Etchemendy sought to reject Tarski's definition by offering examples of intuitive consequences that are not Tarskian consequences (undergeneration) and Tarskian consequences that are not intuitive consequences (overgeneration). And when the two coincide, he argues, it is a matter of accident.[1] We will have more to say about these arguments in Chapter 1 and Chapter 9.

---

[1] The failure of fit between intuitive and Tarskian consequence is the result, Etchemendy believes, of a number of errors made by Tarski, most notably a simple modal error, 'Tarski's fallacy' (see Etchemendy 1990, pp. 85–94).

One True Logic: A Monist Manifesto. Owen Griffiths and A.C. Paseau, Oxford University Press.
© Owen Griffiths and A.C. Paseau 2022. DOI: 10.1093/oso/9780198829713.001.0001

Our aim is not Etchemendy's, however. We largely accept Tarski's definitions but explore their philosophical underpinning. In particular, we will explore the 'common usage' of logical consequence and the features Tarski successfully captures. We will then consider the upshot of this discussion for some of the central questions in philosophical logic. Most importantly, the standpoint from which we view logic should radically change: the finitary perspective adopted by most logicians should change. Logic is a highly infinitary enterprise. We will work towards articulating the *one true logic* of the title, arguing that its infinitary nature is crucial. The plan is as follows.

Summary

Overall, we will first establish that there is one true logic. We will do this by considering some themes in the history of logic and offering a general argument against logical pluralism. Having established the truth of monism in Part I, we work towards finding the *one true logic* in Part II. The historical lessons show that we cannot chase all features that have been taken as characteristic. Instead, we seek an explication of the thought that logic is *formal*. We will explore what it means to call logic formal and settle on an understanding in terms of topic-neutrality. The isomorphism-invariance account is best suited to this task and leads naturally to the claim that the one true logic is highly, indeed maximally, infinitary. We bolster this conclusion with arguments not premised on the nature of the logical constants or relations. In Part III, we then defend the thesis that logic is maximally infinitary against its major criticisms.

Part I

Chapter 1: Conceptions of Logical Consequence

We begin by considering the intuitive features of logical consequence. Virtually everyone accepts that logical consequence involves truth-preservation and some further properties. What these further properties are is the subject of much disagreement. The most prominent ones are *necessity* and *formality*. We will argue, however, that there is significant disagreement about the importance of these properties for logical consequence. The disagreement only multiplies when we consider other features such as normativity, relevance, or *a priori* knowability.

Consider the argument 'Amy is taller than Ben; *so* Ben is shorter than Amy'. This is truth-preserving and it appears to be so as a matter of metaphysical and semantic necessity. Nevertheless, the entailment doesn't seem to be in virtue of *logical form* since the crucial vocabulary—'taller than' and 'shorter than'—is not usually taken to be *logical*. We agree, since we take the form '*x* is taller than *y*; *so y* is shorter than *x*' to be non-logical. If we take necessary truth-preservation to be sufficient for validity, however, we will judge the argument to be valid. As such, the features are in tension, so any attempt to respect all of them must fail. Of course, the situation only worsens as we consider other features like normativity, *a priori* knowability, etc.

In particular, the model-theoretic definition cannot capture all of these features, so we follow Tarski in arguing that it should be seen as an *explication* of the thought that logic is distinctively *formal*. We then tidy this notion of formality up into *topic-neutrality*. A crucial component here is an account of the logical constants, which are the topic-neutral expressions.

Chapter 2: What Is Monism?

The intuitive concept of logical consequence has many different, incompatible, strands. One reaction to this situation is *logical pluralism*: roughly, the pluralist endorses different logics as capturing different precisifications of the rough intuitive conception. In this chapter, we define logical pluralism and its contrary *logical monism*.

The target notion is logical consequence in meaningful discourse and its possible extensions. But the model-theoretic definition is of course defined for formal languages. A crucial component of any account of logical consequence is therefore *formalization*: the process by which we move between meaningful and formal (meaningless) sentences and arguments. We define a logic as a *true* logic, roughly, when formalizations into it capture all and only consequences that obtain among meaningful sentences.

Logical monists claim that there is *one* true logic. Logical pluralists claim that there are *many*. We define logical pluralism more precisely as the claim that at least two logics provide extensionally different but equally acceptable accounts of consequence between meaningful statements. Logical monism, in contrast, claims that a single logic provides this account.

Chapter 3: Against Pluralism

Having explained what logical pluralism is, we argue that any such version will fail. We take Beall and Restall's version and Shapiro's version to be the two most promising ones and offer a general argument against both of them.

We offer a *metalogical* argument against any version of pluralism proper. If a logical pluralist is to engage in metalogical reasoning, what logic should they use? By their own lights, pluralists want their metalogical reasoning to be acceptable in all logics endorsed. If, like Shapiro, the pluralist endorses a great many logics, then logical *nihilism* threatens: very few, if any, inference rules will be allowed in all the logics.

One response would be to restrict the number of logics endorsed, just to classical and intuitionist, say. Here the metalogic may be perfectly acceptable. This is roughly the line taken by Beall and Restall. But it rests on an intuitive picture of consequence with certain 'settled' and other 'unsettled' features of consequence and, as our discussion in Chapter 1 will show, this picture is untenable: at most, truth-preservation is settled.

If logical pluralism is incorrect, we must try to find one logic capturing the intuitive features of consequence. No logic can capture all of them, and out best bet is to pursue formality. What is the one true logic to which formality leads us?

Part II

Chapter 4: The L∞G∞S Hypothesis

The one true logic we will defend as the best explication of the formality of logic is infinitary. In this chapter, we introduce that hypothesis. Very roughly, the one true logic extends finitary logic by allowing disjunctions and conjunctions over any number of formulas, and allowing quantification over any number of argument places. This claim is the *L∞G∞S Hypothesis*, pronounced 'logos hypothesis' and written with infinity symbols replacing the Os for reasons that will be explained in that chapter. We will spend the rest of Part II defending it.

Chapter 5: Beyond the Finitary

The aim of logic understood in a foundational sense is to underwrite the validity or invalidity of arguments in a cleaned-up extension of natural language, including all its technical portions. In this chapter, we offer some 'bottom-up' arguments from this point about logic's motivation to the L∞G∞S Hypothesis.

First, consider argument $\mathcal{A}$: 'There is at least one planet, there are least two planets,..., there are at least n planets,...; *so* there are infinitely many planets'. If we accept—as we do—the validity of this argument in English, then logic's core motivation pushes us beyond first-order logic, which cannot capture the validity of $\mathcal{A}$. We respond to Quine's concurrence-of-ideas defence of first-order logic, based on the concurrence of model- and proof-theoretic ideas via soundness and completeness.

This bottom-up argument pushes us beyond first-order logic. But of course second-order logic can capture $\mathcal{A}$'s validity, so why go as far as a maximally infinitary logic? We marshal some more general arguments which push us beyond second-order resources.

Chapter 6: Isomorphism Invariance

In this chapter, we present the 'top-down' case for the L∞G∞S Hypothesis. The formality of logic—understood as topic-neutrality—relies on an account of the logical constants or logical extensions. The most plausible account of the logical extensions, we argue, is given by isomorphism invariance. In so arguing, we follow and generalize previous ideas of Alfred Tarski and their refinement by Gila Sher. If we take isomorphism invariance seriously, we must endorse a logic of maximally infinitary sort.

We present McGee's Theorem which states, roughly, that the isomorphism-invariant operations in domains of a given cardinality are just those definable in a maximally infinitary logic. We take it to be a necessary condition on the one true logic that, for every isomorphism-invariant relation or operation over its models, the logic contain an expression whose interpretation is fixed as precisely that relation. In other words, the one true logic must contain a logical constant denoting each such isomorphism-invariant extension. This gives us the 'top-down' argument for the L∞G∞S Hypothesis. We conclude the chapter with a proof of McGee's Theorem simpler than McGee's own.

Chapter 7: Towards the One True Logic

This short chapter summarizes our case for the L∞G∞S Hypothesis and ties up three loose ends. The first is what else we know about the one true logic other than that it is highly infinitary. The second is our account's dependence on a theory of models. The third is our account's mathematical consequences, if any.

Part III

In Part III, we defend the L∞G∞S Hypothesis against various objections.

Chapter 8: The Heterogeneity Objection

Invariantism is an idea which can be spelt out in different ways. In this chapter, we consider three different ways to our preferred one, isomorphism invariance. All three are motivated by a common objection to isomorphism invariance. This is the objection that some of the relations or operations isomorphism invariantists take to be logical are too heterogeneous to truly be logical.

The first variant is uniform-isomorphism invariance. This approach takes a relation or operation to be logical if it is the 'same' on domains of different cardinalities. Uniform-isomorphism invariance, combined with logic's closure under definability, also implies that logic is maximally infinitary. From the perspective of this book, then, either uniform-isomorphism invariance or isomorphism invariance will do. Nevertheless, we assess the pros and cons of uniform-isomorphism invariance.

The second variant is Solomon Feferman's strong-homomorphism invariance and the third Denis Bonnay's potential-isomorphism invariance. Neither of these variants has the consequence that logic is maximally infinitary. We argue that they are both untenable.

Chapter 9: The Overgeneration Objection

The most important objection to isomorphism invariance is that it *overgenerates* by finding too many logical constants. In particular, if we accept isomorphism invariance, there will be sentences that are logically true iff the Continuum Hypothesis (CH) is true and others that are logically true iff CH is false. This situation, the objection goes, is intolerable since logical truth is rendered, in some sense, *sensitive* to mathematical truth. But logic isn't mathematics.

We quite agree that logic isn't mathematics; the former is topic-neutral whereas the latter isn't. We argue that it is vital to tread carefully here and consider several precise versions of the objection. In every case, we find that the argument is either unsound, or sound but with an unproblematic conclusion. Crucially, on no sound version of the argument do we find CH—or any other distinctively mathematical claim—being deemed *logically* true, if true; or *logically* false, if false.

Arguments of this sort have often been put forward in the literature against the status of second-order logic as logic. Our arguments can therefore be used by friends of second-order logic in defence of their view.

Chapter 10: The Absoluteness Objection

We consider the *absoluteness* objection, which claims roughly that the set-theoretical notions invoked by the isomorphism invariantist are not robust. We admit this but respond by undermining the motivations for robustness. In particular, we will undermine meaning-theoretic, anti-realist, and independence motivations.

Chapter 11: The Intensional Objection

The final objection we turn to is *intensional*. Isomorphism invariance—whatever guise it takes—offers a purely *extensional* approach to the logical constants, it is claimed. As such, anything coextensive with a logical constant is a logical constant. But this seems wrong: after all, we can cook up all sorts of expressions—like McGee's *unicorn negation*—which are coextensive with a logical constant but which we wouldn't want to call logical.

This criticism is not an objection to our account, which rests on an invariantist account of logical relations but does not identify the logical constants with all and those expressions denoting logical relations. Still, we consider some responses to it. We review the main isomorphism-invariant approaches to the logical constants and show which are more promising than others. Inspired by Gil Sagi, we also show that any overgeneration worry based on unicorn negation and the like is misplaced.

Prologue

Before we begin our arguments for the one true logic, it will be helpful to settle some terminology and state our background assumptions. We first outline our conception of the target phenomenon: natural language. We then explain how we capture this phenomenon in terms of a precise logical-consequence relation. Finally, we comment on the methodology by which natural-language inference and logical consequence should be compared.

Applied logic

Logic, observed Quine, has been a great subject since 1879.[2] It is now also a diverse one, with an infinite variety of logics on offer to suit every logician's taste. Amongst this bewildering assortment of logics, the monist takes a single one to be correct. Monists do not, of course, deny the existence of other logics, for how could they? They maintain, rather, that a single logic captures consequence between meaningful statements. All logics are mathematical systems with their own consequence relations. Yet according to the monist, only one of these systems captures the real logical-consequence relation.

We may draw an analogy between foundational logic—the logic which aims to capture natural-language consequence—and physical geometry. Pure mathematics investigates many different geometries: Euclidean planar, three-dimensional, and higher-dimensional geometry; spherical geometry; hyperbolic geometry; and many more. Theoretical physics/applied mathematics is concerned with a small subset of these: the geometries compatible with physical laws. Indeed, many physicists wish to know how physical spacetime is actually structured, not just how it could have been structured compatibly with the physical laws. One can ask which of the many available pure geometries models that of actual space or spacetime, or which range of geometries does so if more than one does.[3] As has been recognized

---

[2] Quine (1952, p. vii).

[3] Why could there be different geometries of the one spacetime? Different choices of primitives give strictly different mathematical systems. Moreover, the mathematics used to describe spacetime could pack in extra structure than we take spacetime to have; e.g. if a proposed geometry of our actual spacetime features a metric d then presumably using a metric $d^* = 2d$ would not correspond to a physical difference. For more general discussion of 'gauge theories'—physical theories that exhibit excess structure—see Weatherall (2016).

One True Logic: A Monist Manifesto. Owen Griffiths and A.C. Paseau, Oxford University Press.
© Owen Griffiths and A.C. Paseau 2022. DOI: 10.1093/oso/9780198829713.001.0002

since Einstein (1921), indeed for decades prior, these are questions for physics rather than mathematics.

Similarly, what we might call the *pure* logician investigates a variety of logics. A typical pure logician might find their natural home in a mathematics department. In contrast, the *applied* logician wants to know what the logic or logics of language is/are. A typical applied logician is more likely to be housed in a philosophy department. Now this analogy is not supposed to be anything more than that. We appreciate for instance that some applied logicians are more interested in what the logic of an ideal language should be than in the logic of our actual language. The example of Frege, whose interest was in a language better suited than ours to the purposes of science, springs to mind. And in no serious sense are physicists interested in what spacetime *should be* as opposed to what it *is*. We also appreciate that the geometry(ies) of physical space might be quite different from the geometries studied in pure mathematics. Such a geometry is for instance warped by the massive objects it contains, whereas the spaces studied by pure mathematicians typically do not contain massive objects. Still, the gist of the analogy and our purpose in introducing it should be clear enough: this book is a work of applied and not pure logic. Our task is much more akin to the mathematical physicist's than the pure mathematician's.

Something like the distinction between pure and applied logic is commonplace in the literature, though different authors draw it in different ways. Priest (chs 10 & 12 of 2006, 2001), for example, also likens logic to geometry and takes any application of logic—to electronic circuitry say—to be a form of applied logic in the relevant sense. As he sees it, the application of logic to reasoning is its 'canonical application'.[4] He writes that logic is in the business of determining 'what follows from what—what premises support what conclusion—and why' (2006, p. 196), something we discuss further in Chapter 3. Here we focus on the application of logic to implicational relations between natural-language premises and conclusion. These relations, for us, are what we try to get right when we engage in deductive inference. We are less interested in non-logical implication, and consequently in other types of inference, such as inductive or abductive.

As just hinted, we take capturing implication and capturing reasoning as distinct applications. The implications of some premises are their logical consequences; they follow from them, whether or not one can deduce them from the premises. In contrast, an inference is what an agent does when she deduces a conclusion from some premises. Reasoning or inference tries to respect implication, though is distinct from it. Thus we write 'implicational' rather than 'inferential' whenever we are interested in what follows from what—as we typically will be—rather than what

---

[4] Priest contrasts this with logic's application to electronic circuitry. Although even here one might think that what is being modelled are logical relations between propositions about circuits.

can be deduced from what.[5] All this applies even to idealized notions of reasoning (which for example prescind from human subjects' errors in reasoning).[6]

It follows in particular that there is no reason to suppose at the outset that the correct foundational logic is completable by a (sound and effective) deductive system; or, if one is a pluralist, that the correct foundational logics are each completable by such a system. Perhaps no deductive system can capture all logical entailments. (Incidentally, we use the words 'implication', 'consequence', and 'entailment' interchangeably, with the epithet 'logical' understood when omitted.)[7] Implication is modelled by model-theoretic consequence ($\vDash$) and derivability by deductive consequence ($\vdash$). On this standard, model-theoretic understanding, there is no reason to suppose that deductive consequence exhausts logical consequence. We shall consider and reject a reason for thinking that it does in §5.2. In arguing for an infinitary true logic in Part II of the book, we will definitively establish that logical consequence outstrips derivability.

From E to E_c and beyond

Our starting point is meaningful discourse, typically couched in natural language.[8] Let E be a generic language consisting of meaningful sentences. For concreteness, there is no harm in identifying E with English, as it's the language this book is written in. E includes everyday as well as technical vocabulary, such as the vocabulary of mathematics and science. Importantly, E's sentences may also be ambiguous.[9] 'John likes American English speakers' could for instance mean that John likes speakers of the American dialect of English, or that he likes speakers of the English language who are American. It follows from the first but not the second reading that John likes Juan, a Panamanian speaker of American English. The logical properties of the first disambiguated sentence are different from the second's. Consequently, to determine the logical properties of the sentence 'John likes American English speakers' requires prising the two readings apart.

Logicians are often not very interested in the logical relations of a language such as E. Their focus instead is on *cleaned-up* natural language. We may denote

---

[5] For more on the distinction between implication on the one hand and reasoning on the other, see Harman (1986, ch.1), succinctly summarized in Harman (2009), and Steinberger (2019).

[6] So long as the sense of reasoning/inference is not *so* idealized that it means nothing other than the ability to accurately reflect implicational facts.

[7] With the usual act/outcome ambiguity, resolvable by context; e.g. 'entailment' can mean the relation that holds between some premises and a conclusion, or the conclusion itself.

[8] The term 'natural language' in this context is not ideal. Constructed languages, such as Esperanto and a bevy of lesser-known ones, are not generally regarded as natural, even if they are like 'natural languages' such as English in all respects that matter to us here.

[9] Guided by etymology, some people call a sentence *ambiguous* if it has two meanings and *polysemous* if it has two or more meanings. We use 'ambiguous' in the latter, broader sense.

cleaned-up English, our representative natural language, by E_c (c suggests 'cleaned-up'). What is it for a language to be a cleaned-up version of another? At a minimum, the former should resolve all the latter's structural and lexical ambiguities. Thus E_c should be thought of as English purged of ambiguity. That is how we proceed in this book, even if we carry on using English sentences as proxies for statements of E_c, that is, statements of disambiguated English.

Some authors would add further requirements. One such might be that E_c's sentences be free from both ambiguity *and* vagueness.[10] That stipulation, however, leaves no room for forms of monism and pluralism motivated by accounting for vagueness. Fuzzy logic, supervaluations, multi-valued logic, and intuitionistic logic—not to mention classical logic—have all been proposed as the right logic for vague languages. Whether or not the logic of vague statements is non-classical, it would be a mistake to rule such logics out by fiat. We want to make definitional room for, say, fuzzy-logic monism. That said, our decision not to purge E_c of vagueness will have little impact on our discussion.

Applied logicians thus wish to give an account of logical relations among sentences of at least E_c, a cleaned-up version of E (English). Just how cleaned-up E_c should be is contentious; to preserve generality in Part I, we take the minimal reading of E_c as E shorn of ambiguity. A stronger reading would risk loading the dice in favour of monism, since pluralists might argue that we can clean up language in distinct ways and that there is no unique E_c in the first place. Monists thus believe that the account of E_c's logical properties can be given in terms of a single logic, whereas pluralists demur. But before mapping out their disagreement more precisely, we must ask whether it is really E_c that applied logicians are interested in, or an extension of E_c.

Logicians these days usually publish in English, the academic community's *lingua franca* as well as the world's more generally. It would, however, be hopelessly parochial for the applied logician to give an account of consequence that applies solely to English.[11] For one thing, there is nothing special about English from a logical point of view. Logicians should be, and often are, just as interested in other natural languages. For another, there are plenty of words in other languages that have no equivalent, or at least no exact equivalent, in English. The usual stock of examples includes German's *Schadenfreude*, Italian's *omertà*, and Danish's *hygge*. Musing on whether these words can be paraphrased in English is a diverting pastime for bilingual speakers. But the verdict on this score does not affect the fact that English lacks concepts expressed by some other languages, a point sufficiently

---

[10] Thus Stewart Shapiro: 'For convenience, we'll take logical consequence to relate sentences in interpreted languages, free from the usual ambiguities, indexicals, vagueness, and the like' (Shapiro 2014, p. 19). Since ambiguity, indexicality, and vagueness are all distinct phenomena, it is not clear what Shapiro means by 'and the like'. For a little more on indexicality, see below.

[11] Or more precisely, to a cleaned-up version of English. We often omit this qualification for brevity.

obvious that we do not pause to argue for it. English is not, in this sense, a universal language.

A more catholic choice of target language for the applied logician would be the union of all existing natural languages (including their scientific branches). To be more inclusive still, we could throw in all extinct languages, for example Proto-Indo-European. Such a choice would still be far too parochial, however; we need to widen our horizons further. Natural languages accrue words by the day, and those without a written record shed them too. In a few years, English will have acquired sentences it can't express today, just as the sentence 'I can't play my DVD on my laptop' would have been incomprehensible to Victorians. As applied logicians ourselves, we are not interested in giving an account of logical consequence that is valid merely for the precise moment this book was completed or is being read. Furthermore, even if natural languages were set in stone and no longer subject to change, they *could* have been different from what they are. Just as a meteorite crashing on Earth in 1901 might have prevented us from ever adding the words 'DVD' and 'laptop' to English, so future natural languages could be different from how they in fact turn out.

We are edging closer to the idea that the applied logician's interest should not be in any particular natural language or its cleaned-up version (E or E_c), but in these languages' extensions. A thought experiment helps cement this conclusion. Suppose that in the future we come to discover a planet in another galaxy on which intelligent life exists. As our astronauts take their first steps on the planet with trepidation, the first being they come across is someone we subsequently call 'Aderph'. Aderph is the first of the many zoovarks we encounter on the planet, later christened 'Aderphia'; 'zoovark' is the name we coin for the creatures inhabiting Aderphia. A host of other Aderphia-related names subsequently enter English. And with these names' incorporation, some new arguments are expressible in the expanded language, such as: 'Aderph is a zoovark; all zoovarks are friendly; therefore, Aderph is friendly'. Obviously enough, this last argument is valid.[12]

In sum, a complete account of logical consequence takes in more than current natural languages. We may thus distinguish two projects for the applied logician:

---

[12] Someone might object that 'Aderph' and 'zoovark' are already expressible in English, even if the words themselves are not currently employed. 'Aderph' means *the first being we encounter on the first planet other than Earth on which humans discover intelligent life*; and 'zoovark' means *member of the species on the first planet other than Earth on which humans discover intelligent life*. As is familiar from the vast literature on descriptivism, such descriptions, or variations on them, are not very plausible candidates for the meanings of the words 'Aderph' and 'zoovark'. In any case, staunch descriptivists may use as an alternative example expressions in other current languages lacking exact counterparts in English. They should also recognize that some concepts are not expressible in any current language. They should do so even if they maintain that no example can currently be given of an inexpressible thought, on pain of expressing the currently inexpressible.

Parochial Project: account for logical consequence among E_c-sentences.

Ambitious Project: account for logical consequence among sentences of extensions of E_c.

Although our first label is pejorative, we hope to have explained our reasons for it. Turning to the ambitious project, we immediately face a tradeoff between the account's generality and our confidence in its deliverances. In its most ambitious form, it accounts for logical consequence among sentences of *all* possible extensions of E_c. But such ambition is overweening. Neither of the present authors has a clear idea of what all possible extensions of E_c are, and we suspect no one else does either. We don't even know whether it is coherent to think of all of E_c's possible extensions as a completed totality. Our grip on the notion of any possible meaningful language is too weak to pursue the most ambitious version of the Ambitious Project.

A more modest version of the Ambitious Project holds out more promise. By moving to an extension of cleaned-up English, we go beyond a narrow concern with our contingent linguistic inheritance. But staying within touching distance of actual natural languages allows for reasonably informed debate. The kind of extensions of E_c that will occupy us in Part II are precisely of this kind. For example, English augmented by Aderphian words such as 'zoovark' clearly falls under the scope of our discussion. Similarly, if $E_c(\kappa)$ is cleaned-up English augmented with a name for the cardinal κ (and its obvious syntactical rules), then each of the languages $E_c(\kappa)$ is also within our remit.[13] With these in hand, we can then define the cardinality quantifiers 'There are κ', which we shall argue in Part II are all logical.

Let's adopt, then, the label E_c^+ for a (cleaned-up) language that extends cleaned-up English. The subscript 'c' denotes 'cleaned up' as earlier, and the superscript '+' reminds us that we're concerned with an extension of E_c. As just noted, such an extension should not leave English too far behind, if reasoned debate is to be had. All that said, Part II's conclusions about the contours of the one true logic are radical enough *even when restricted to E_c*.

In summary, giving an account of logical consequence for (cleaned-up) current natural languages is an interesting, though parochial, project. Philosophers of logic should, and implicitly do, aim a bit higher. They would in principle like to take in all meaningful language. Yet to get informed debate off the ground, they confine themselves to extensions of current natural languages we have a reasonable grip on. Epistemic humility counsels against aiming too high and trying to give an account of *all* language whatsoever, whatever exactly that might mean.[14]

---

[13] Clearly, for all the values of κ definable in English—for example $\kappa = \aleph_0$ or $\kappa = n$ for finite n—$E_c(\kappa)$ is just E_c. However, since English only defines countably many such κ, all but countably many of the languages $E_c(\kappa)$ properly extend E_c. See Chapter 5 for more.

[14] For all we know and have said, the one true logic outlined in Part II *may* govern all language.

Relata

We say that a natural-language argument consists of a set of premises and a conclusion. A special case is when the premise set is empty: natural-language statements entailed by the empty set of premises are known as *logical truths*. Another special case is when the conclusion is missing. If a natural-language argument with an empty conclusion is valid then its premises are semantically inconsistent.[15] If $\mathcal{L}$ is a logic (defined below), we call any $\mathcal{L}$-sentence that is true in all interpretations an $\mathcal{L}$-*validity*.[16] If $\mathcal{L}$ is first-order logic, then the (uninterpreted) formula $\forall x(x = x)$ is a logical validity and should not be confused with the logical truth 'Everything is self-identical'. (The previous sentence illustrates a presentational point: we will often drop quotation marks, for example when citing logical formulas, when no confusion is likely to result. To avoid clutter, we also avoid the use of Quine quotes.)

Thinking of the premises of a natural-language argument as a set is useful and conventional in logic, if a little odd-sounding to non-logicians. But it isn't obligatory. Set talk may for instance be replaced by a plural idiom: we may say that *some premises* imply a conclusion as opposed to saying that *a set of premises* implies a conclusion. In an equally dissenting vein, multiple-conclusion logicians take an argument to have a *set of conclusions* as well as a set of premises.[17] Others might take consequence as a relation between multi-sets of premises (allowing repetition of premises) and a conclusion (Restall 2000). Another approach might be to think that the *order* of the premises is important, though nobody to our knowledge has endorsed this. Nothing bars any of these from being logical monists. We shall assume the standard picture of a set of premises implying a single conclusion, without much commitment to its correctness. It is consistent with our perspective in Part I, and with most of our discussion in the rest of the book, that a difference in view about the relata of the consequence relation might result in a difference in logic.[18]

Our talk of premises and conclusion so far has been silent on their nature. Here the usual candidates are sentences, statements, or propositions.[19] Again, we believe that any choice here can be made consistent with our arguments, but we will assume that natural-language premises and conclusions are *statements*.

What are statements? A popular, though by no means universal, approach is this. A statement is an ordered pair consisting of an E_c^+-declarative sentence-type (e.g. 'I am cold') and a context (e.g. the North Pole at noon on the 1st of January

---

[15] The proof-theoretic notion is syntactic inconsistency. The qualification that an argument's set of premises may be empty or that it has no conclusion is implicit in all that follows.

[16] In the proof-theoretic variant, the notion of an $\mathcal{L}$-theorem is akin to that of an $\mathcal{L}$-validity, an $\mathcal{L}$-theorem being provable from the empty set of assumptions.

[17] The classic reference remains Shoesmith and Smiley (1978).

[18] A fairly trivial way in which this happens is that multiple conclusions are sometimes understood disjunctively, sometimes conjunctively, which is not an ambiguity in the object language.

[19] See Gillian Russell (2008) for a form of logical pluralism that trades on this distinction.

2001). If $\langle s, c \rangle$ is a statement, c includes all the relevant information for determining the truth-value of the sentence s. A statement $\langle s, c \rangle$ is then true iff an utterance of s in c expresses a true proposition. Although this account suggests that sentences are utterance-types, we shall think of them *either* as inscription-types *or* as utterance-types depending on the needs at hand.

Inference applies to statements thus understood, as long as context remains fixed. For example, the argument 'I'm cold and tired; *so* I'm cold' is pre-theoretically valid, as long as the speaker who utters the premise is the same as the speaker who utters the conclusion. This is the standard approach: see Quine (1982, p. 56) for an influential statement, and Rumfitt (2015, p. 33) for a more recent one. A restriction of this usual approach would be to consider logical consequence as applying only to context-insensitive statements. In the other direction, applied logicians could be more ambitious and consider logical relations between statements in different contexts.[20] But as we say, any reasonable choice here is compatible with all this book's arguments and claims, give or take a reformulation or two.

We will be interested throughout in statements made by declarative sentences.[21] Erotetic logic (the logic of questions), imperative logic (the logic of imperatives), and any other logics of non-declarative sentences are all assumed to be in some sense dependent on the logic of declarative sentences.

In summary, the target phenomenon of the applied logician is the relation of logical consequence in cleaned-up extensions of natural language. This we take to hold between a (possibly empty) set of statements as premises and a single statement as conclusion. That said, the difference between statements and sentences will matter little, and we shall frequently speak of the two interchangeably. More importantly for us, natural languages must be tidied up (from E to E_C) and, most importantly, the applied logician should be concerned with their possible extensions (from E_C to E_c^+).

Monism

We argue that there is *one true logic*. To understand this slogan, we must first know what a *logic* is and then have some counting principle to know exactly when there is just one such. First, we take a *logic* to be a formal language (consisting of a vocabulary and a grammar) and a semantics assigning meanings to the formal expressions and ultimately defining *logical consequence*. A logic may also come with a deductive system (defining *deducibility*); but since the focus of this book is model-theoretic, we will have to little to say about deductive systems.

---

[20] See Iacona (2010) for discussion of cross-context notions of validity.

[21] This assumption isn't uncontroversial; e.g. given the right context, tone and so on, a statement such as 'The door stays shut.' might be taken by everyone as imperative, even though it is grammatically declarative. Our points are unaffected by the existence of these sorts of cases.

Our central concern is with the semantic notion of *logical consequence*, which holds between some set of statements S and a sentence s just when any model of S is a model of s. We use 'S logically entails s', 'S logically implies s', 's is a logical consequence of S' and 'the argument with premise set S and conclusion s is logically valid' interchangeably; and we occasionally drop the adverb 'logically', as mentioned. We will also discuss logical truth, defined in terms of logical consequence. Other logical properties, such as consistency and independence, can also be defined in terms of logical consequence, but we will generally have less to say about them.

The aim of the applied logician is to determine the consequence relation that best matches the logical-consequence relation in natural language, or the consequence relations that do so if there are several of them. A logic's consequence relation is therefore its most important feature for our project and, for that reason, we will often *identify* a logic with its consequence relation. From this perspective, we can say that two logics are identical just if their consequence relations are identical, i.e. coextensive.

This isn't quite enough, however. There are various ways in which, by this standard, two logics might be deemed non-identical even though the difference is trivial or notational. For example, if two logics are both classical propositional logics, but one takes its atomic letters from the English alphabet and the other from the Greek alphabet, then their consequence relations will not be identical. Or if one logic takes its atomic sentences from lower-case letters of the English alphabet and the other from upper-case letters, they too will be judged non-identical. But plainly these differences are trivial and we don't think that logical pluralism can be had so readily.

Indeed, when we come to define pluralism in Part I, we will see that such trivial differences do not suffice for pluralism. As a first pass, we'll say that two logics are identical just when their consequence relations—most simply understood as a set of ordered pairs of premise sets and conclusion—are *isomorphic*. In this way, the English/Greek alphabet logics are the same, since it is a trivial matter to transform one of their consequence relations into the other. And similarly for the lower-/upper-case logics. But the consequence relations of, say, classical first-order logic and intuitionistic first-order logic are distinct, which is what we want.

The method

We've now seen that the task of the applied logician on which we are embarked is to determine the logic (or logics) that best matches (match) that (those) of cleaned-up and possibly extended natural language. What is meant here by *match*?

A strong answer is that some logic's consequence relation provides a *conceptual analysis* of the target phenomenon. The thought here is that what it *means* for a statement to follow logically from some set of statements in natural language is given by the logical consequence relation of some logic. Whatever this amounts to, it must entail that the two are *coextensive*, since meaning determines extension.

There are immediate problems here. First, the extension of inference in natural language will consist of (sets of n-tuples of) *natural-language* statements, and that of logical consequence in some logic will consist of (sets of n-tuples of) *formal-language* sentences. However, natural and formal languages are distinct, so the claim of coextension is doomed to failure. To fix this, as we did when giving the identity criteria of logics, we will appeal to the notion of an isomorphism: the extensions of the two concepts are isomorphic. How are we to translate between formal and natural languages in order to judge whether such isomorphisms exist? We need a notion of formalization, a crucial but little-discussed notion we will return to in Chapter 2.

Second, the extension of some formal consequence relation is precise and it is far from clear that the extension of logical consequence in natural language is precise. Indeed, a major theme of Chapter 1 will be the messiness of this notion. But if a logic's consequence relation has a precise extension and that of natural-language inference is fuzzy, they can never be isomorphic.

There are, however, attempts to show a precise concept to be extensive with an (apparently) vague one, namely *squeezing arguments*. The original squeezing argument is offered by Georg Kreisel (1967) and concerns notions of logical consequence (though, crucially, Kreisel's doesn't involve any natural-language relation).[22]

The general strategy in a squeezing argument is to take some possibly vague, informal concept I (this could be *any* informal concept—it need not be the concept of logical consequence) and sandwich it between two precisely defined concepts P_1 and P_2. The first step is to show that, in spite of its imprecision, falling under I is clearly *sufficient* for falling under P_1:

(1) $I \rightarrow P_1$

The second step is to find another formal concept P_2 such that, in spite of the imprecision of I, falling under P_2 is clearly necessary for falling under I:

(2) $P_2 \rightarrow I$

The third, and final, step is to establish a result about the precise concepts:

(3) $P_1 \rightarrow P_2$

---

[22] More recently, Hartry Field (2008, pp. 47–8) offers a squeezing argument on notions of logical consequence. See Peter Smith (2011) for criticism.

(3) squeezes the sandwich together so that all three concepts I, P_1, and P_2 must all have the same extension. And, because P_1 and P_2 have precise extensions, I must have a precise extension too.

Usually, for this sort of argument to have any hope of success, the imprecise notion must still be *quite* precise.[23] Again, for reasons we give in Chapter 1, we are doubtful that such a notion can be found for natural-language consequence. Second, Griffiths (2014) has argued that, given the notion of formalization needed to get such arguments running, they will inevitably beg the question. For these reasons, we are doubtful that *coextension* is possible to establish.

We should not seek coextension when attempting to account for logical consequence in natural language. What other method is available? Tarski again makes a plausible suggestion when he writes about the 'common concept of consequence':

> Its extension is not sharply bounded and its usage fluctuates. Any attempt to bring into harmony all possible vague, sometimes contradictory, tendencies which are connected with the use of this concept, is certainly doomed to failure. We must reconcile ourselves from the start to the fact that every precise definition of this concept will show arbitrary features to a greater or less degree.
>
> (Tarski 1936, p. 409)

This language does not fit well with a coextension claim, because it admits that the target is too vague and fluctuating in usage to be precisely captured. As such, Tarski would not have claimed that his early model-theoretic notion of consequence was coextensive with the 'common concept'.

Rather, Tarski sought a Carnapian *explication* of the common concept, which is a precise refinement replacing a vague, intuitive conception, narrowed down for certain technical purposes. Carnap wrote that

> The task of making exact a vague or not quite exact concept used in everyday life or in an earlier stage of scientific or logical development, or rather replacing it by a newly constructed, more exact concept, belongs among the most important tasks of logical analysis and logical construction. We call this the task of explicating, or of giving an *explication* for, the earlier concept.   (Carnap 1947, pp. 8–9)

When we have given up on the hope of establishing coextension, explication is an attractive alternative. It involves taking an intuitive notion and replacing it for certain purposes with a precise notion, allowing for some idealization of the target phenomenon.

---

[23] As Smith (2011) argues.

This is the spirit in which we advance our one true logic in this book. We will argue that there is a uniquely correct way to explicate logical consequence in natural language.

Classical Logic

In Part II, we shall argue that the one true logic must be maximally infinitary, in a sense that will be made precise there. In doing so, we shall generally assume that the one true logic's first-order fragment is classical and that its semantics is also classical.[24] We shall also stick to non-modal, non-epistemic, etc. logic. This is for four reasons.

The first is that we wish to focus squarely on this aspect of a logic: that it is maximally infinitary. It would be distracting to rework our arguments from each non-classical perspective, or from the perspective of various modal, epistemic, etc. extensions to first-order classical logic.

It would also be unfeasible to do so (this is the second reason). We would have to consider all varieties of non-classical logics, survey all possible modal, epistemic, etc. logics, and evaluate their pros and cons. This would be the work of several tomes rather than the manifesto-like monograph we intend this book to be. Defending classical logic against all-comers—intuitionistic logic, relevance logic, quantum logic, and new-fangled alternative logics—is too large a task.[25]

The third reason is that our focus is more on the scientific parts of natural language. The tradition we stand in ultimately traces back to Frege's *Begriffsschrift* of 1879, a text whose primary aim was to serve as a methodological basis for the sciences. Other twentieth-century philosopher-logicians such as Carnap, Tarski, and Quine followed in Frege's footsteps. From this perspective, which focuses on the use of logic in broadly scientific contexts, some of the vagaries of natural language may be legitimately ignored. If irreducibly messy natural language is a useful tool which is nevertheless inappropriate for scientific purposes, then the logician need not cause offence to the linguist in setting it to one side. Better for a farmer to spend time sharpening a scythe than a hammer. Or as Frege himself put it in the *Begriffsschrift*, when sharpness is sought, better to use a microscope than an eye. As a corollary, we don't assume that the logical form a formalization ascribes to a sentence is identical to the semantic structure which a theory of meaning ascribes it; semantic structure may or may not be identical to logical form.[26] That

---

[24] As we do not individuate logics by their deductive systems, disputes about which rules logics (such as classical first-order logic) contain are irrelevant here. By a classical semantics, we mean roughly the kind of bivalent semantics whose paradigm is the Tarskian semantics for first-order classical logic, as detailed in most introductions to logic, e.g. Chapter 5 of Halbach (2010).

[25] For a start on how to adjudicate between rival logical systems, see Rumfitt (2015).

[26] For an overview of this aspect of the 'classical conception' of logical form, see Jackson (2006).

said, we take our work to be not just consistent with but complementary to work in linguistics, for example semantic accounts of quantifiers, as showcased in Peters and Westerståhl (2006).

The fourth reason is that classical logic has much to be said for it. There is no good reason to absolve logic from the usual abductive process of justification.[27] Like any scientific theory, our theory of logical consequence must be assessed by its fit with the evidence and the usual scientific virtues. And once we assess logics in this abductive manner, it should be clear that classical logic fares rather well compared to its rivals. Moreover, the opposition to classical logic is fragmented. Its enemies have criticized it on many different fronts, basing their arguments on a diverse array of considerations ranging from quantum phenomena to semantic paradoxes and vagueness. To their advocates, these phenomena suggest different revisions to classical logic. Opponents of classical logic tug in different and incompatible directions, to some extent cancelling each other out, and thereby failing to undermine their target's canonical status. The opposition is divided, so classical logic rules. Furthermore, the study of rivals to classical logics also generally takes place in a classical metatheory. As Timothy Williamson has stressed, those who advance such rivals are either adopting a questionably instrumentalist attitude to the metatheory, or cannot make good on the commitment to use the rival logic as their foundational logic.

All that said, the classical perspective adopted here is not necessary for all our conclusions. We believe that *most* of what we say in the book can be be restated, under suitable reformulation, by *most* non-classical logicians. Many of our arguments for our claim in Part II that the one true logic is maximally infinitary could be recast in a non-classical setting. Non-classical logicians of various (but not all) stripes can adapt our arguments in Part II and convert them into considerations for adopting strong infinitary versions of their preferred logics. Similarly for our arguments in Part III.

---

[27] As Timothy Williamson (2017), and more generally 'anti-exceptionalists', have emphasized.

PART I
MONISM VS PLURALISM

1

Conceptions of Logical Consequence

Logical consequence is the central concept in the philosophy of logic. We will argue that there is no single, well-behaved intuitive conception of consequence. This conclusion will be crucial for what follows, since it will form a key premise of our argument against logical pluralism in Chapter 3. Further, it will inform our articulation of the one true logic in Part II. Our claim about intuitive conceptions of consequence is an empirical one and so this chapter will be a largely historical discussion, albeit an opinionated one. The single most influential work on modern conceptions of consequence is Alfred Tarski's classic 'On the concept of logical consequence' (1936) and that's where we'll start. We'll mention Tarski and Vaught's 1956 article, in which the model-theoretic conception in its full-fledged contemporary guise first emerges, more briefly.

1.1 Tarski on logical consequence

Tarski begins by noting that the concept of logical consequence

> is one whose introduction into the field of strict formal investigation was not a matter of arbitrary decision on the part of this or that investigator; in defining this concept, efforts were made to adhere to the common usage of the language of everyday life.   (Tarski 1936, p. 409)

In other words, the language of logical consequence has a usage in everyday life prior to any attempts to model it. As such, any theory of logical consequence should start with an investigation of this 'everyday' conception of consequence.

But, Tarski goes on, there is a problem here from the start, since this 'common conception' of consequence

> is in no way superior to other concepts of everyday language. Its extension is not sharply bounded and its usage fluctuates. Any attempt to bring into harmony all possible vague, sometimes contradictory, tendencies which are connected with the use of this concept, is certainly doomed to failure. We must reconcile ourselves from the start to the fact that every precise definition of this concept will show arbitrary features to a greater or less degree.   (Tarski 1936, p. 409)

One True Logic: A Monist Manifesto. Owen Griffiths and A.C. Paseau, Oxford University Press.
© Owen Griffiths and A.C. Paseau 2022. DOI: 10.1093/oso/9780198829713.003.0001

Here, Tarski suggests that the task of modelling the common concept of consequence is one of *explication* (though he does not use that label). This is because the concept is 'vague, sometimes contradictory' in its *intension*; for example as we will see, there is debate as to whether it is formal, modal, or normative. As a result, its *extension* is 'not sharply bounded'. An example: is the principle of bivalence a logical law? Or is the law of non-contradiction? We should not, therefore, seek *coextension* when we attempt to model logical consequence, as there is no sharp extension to match.

Having commented on the messy nature of the 'common conception' of consequence, Tarski goes on to survey deductive approaches to logical consequence. These approaches are flawed, he believes, because concepts of deducibility in deductive systems will be ω-incomplete: they will not allow the derivation of a universal generalization A over numbers from all of its numerical instances $A_0, A_1, \ldots, A_n, \ldots$. This is a problem, Tarski maintains, since 'intuitively it seems certain that the universal sentence A follows in the usual sense from the totality of particular sentences $A_0, A_1, \ldots, A_n, \ldots$'. We can, Tarski (1936, p. 412) notes, use Gödelian techniques to form finitary rules that allow the derivation of these universal generalizations from their instances but the resulting theory will still be ω-incomplete:

> In every deductive theory (apart from certain theories of a particularly elementary nature), however much we supplement the ordinary rules of inference by new purely structural rules, it is possible to construct sentences which follow, in the usual sense, from the theorems of this theory, but which nevertheless cannot be proved in this theory on the basis of the accepted rules of inference.
>
> (Tarski 1936, p. 412)

Given this major flaw in deductive approaches, a different, *semantic* approach is needed, and Tarski begins by considering Carnap's suggestion:

> The sentence *X follows logically* from the sentences of the class *K* if and only if the class consisting of all the sentences of *K* and the negation of *X* is contradictory.
>
> (Tarski 1936, p. 414)

Tarski notes that Carnap's notion of 'contradictory' is 'too complicated and special' to discuss in detail. He gives further reasons to reject Carnap's definition, but first begins to sketch his own. We should be concerned, he writes, 'with the concept of logical, that is, *formal*, consequence, and thus with a relation which is to be uniquely determined by the form of the sentences between which it holds' (1936, p. 414).

This is a clear indication that the intuitive notion of consequence that Tarski seeks to explicate is one of consequence *in virtue of form*. (We will consider what this slogan means in much more detail later.) The relation of logical consequence, he adds, 'cannot be affected by replacing the designations of the objects referred to in these sentences by the designations of any other objects' (1936, p. 415). On this basis, he offers his first attempt at a definition:[1]

> **(F)**   If, in the sentences of class K and in the sentence X, the constants—apart from purely logical constants—are replaced by any other constants (like signs being everywhere replaced by like signs), and if we denote the class of sentences thus obtained from K by K', and the sentence obtained from X by X', then the sentence X' must be true provided only that all sentences of the class K' are true. (Tarski 1936, p. 415)

He notes that (F) is a necessary condition for logical consequence but cannot be *sufficient* because the language may be expressively weak, for example, *Fa* could imply *Gb* simply because any object that is F and not G lacks a name in the language. In a footnote, Tarski then points out that expressively weak languages also provide a counterexample to Carnap's definition (1936, p. 416, fn. 1).

The notions of *satisfaction of a sentential function* and of *model* provide Tarski with the solution to this problem. As is familiar from Tarski (1935), *satisfaction* is a semantic relation that holds between objects or sequences of objects and sentential functions (sentences containing free occurrences of variables, which we would today call *open* sentences or formulas). In particular, a sequence of objects satisfies an open sentence just if the result of assigning to each free variable, taken in order, the corresponding object is a true sentence. Tarski's explanation of sentential functions and models warrants extended quotation:

> One of the concepts which can be defined in terms of the concept of satisfaction is the concept of *model*. Let us assume that in the language we are considering certain variables correspond to every extra-logical constant, and in such a way that every sentence becomes a sentential function if the constants in it are replaced by the corresponding variables. Let L be any class of sentences. We replace all extra-logical constants which occur in the sentences belonging to L by corresponding variables, like constants being replaced by like variables, and unlike by unlike. In this way we obtain a class L' of sentential functions.

---

[1] Gómez-Torrente (1996, p. 130) claims that the label (F) here stands for 'formality'. However, Stroińska and Hitchcock note in their English translation of the Polish version of Tarski's paper (Tarski 2002) that in this version Tarski uses (W), abbreviating 'wynikanie', Polish for 'consequence'. Tarski, whose second language in the 1930s was German, also wrote a German version of the paper alongside the Polish one, in which he used the label (F) instead of (W). Stroińska and Hitchcock plausibly argue that (F) in this version abbreviates 'Folgerung', German for 'consequence'. The English translation uses the label (F) because it was the German rather than the Polish version of the article that was later used as the basis of the only twentieth-century English translation.

> An arbitrary sequence of objects which satisfies every sentential function of the
> class *L'* will be called a *model* or *realization of the class L of sentences*
>
> (Tarski 1936, pp. 416–17)

For Tarski, then, we form a sentence's sentential function by keeping the logical constants unchanged and replacing the non-logical constants uniformly by
variables (same non-logical constants by the same variable, distinct nonlongical
constants by distinct variables) of the appropriate sort. A Tarskian model of a
set of sentences is a sequence of objects that satisfies the sentential function of
each member of that set. He is then in a position to put forward the following
definition:

> The sentence *X* follows logically from the sentences of the class *K* if and only if
> every model of the class *K* is also a model of the sentence *X*.
>
> (Tarski 1936, p. 417)

So *X* is a logical consequence of *K* just if every sequence of objects that satisfies
the sentential functions of the sentences in *K* also satisfies the sentential function
of *X*. And a sentence is a logical truth just if every sequence of objects satisfies its
sentential function.[2]

Tarski states that this definition 'agrees quite well with common usage', captures
the necessary condition (F), and avoids the problem of expressive weakness. He
ends by noting that a hole remains in his account: the notion of *sentential function*
makes use of the notion of *logical constant*, and he has given no demarcation of the
logical constants. He leaves this demarcation problem to future research but also
considers 'it to be quite possible that investigations will bring no positive results
in this direction, so that we shall be compelled to regard such concepts as "logical
consequence" ... as relative concepts' (1936, p. 420).

In sum, in his definition of logical consequence and validity, Tarski applies the
theory of truth and satisfaction he had developed in the early 1930s. He replaces
predicate and individual constants with variables and defines the conclusion to
be a logical conclusion of the premises just when the former satisfies all variable
assignments satisfied by the latter. Tarski called these variable assignments 'models', although the model-theoretic definition in its more precise contemporary
form emerges a little later, in Tarski and Vaught (1956).

---

[2] We may think that a sentence is a logical truth for Tarski just if the corresponding *universal sentence*
is true, i.e. if the result of putting appropriate universal quantifiers before its sentential function is true.
This isn't quite correct, however, since the universal quantifier can only be understood relative to a
domain, whereas in Tarski's definition we are to consider *all* sequences of objects.

1.2 Accounts of logical consequence

Tarski's 1936 paper helpfully distinguishes three issues. The first is the issue of the intuitive conception of logical consequence. As we saw, Tarski notes that there are many aspects to the intuitive notion and he chooses to focus on *formality*. There is a further question about the extent to which Tarski was also interested in a modal notion of consequence, a question we will here bracket. What is clear is that formality is the central plank of his account.[3]

The second issue is the precise definition of logical consequence. Tarski, as we saw, considers four precise definitions: the deductive account, Carnap's semantic account, his own (F) and the model-theoretic definition on which he settles.

The third issue is the relationship between the intuitive conception and the precise definition. Given the vague, fluctuating boundaries of intuitive consequence, we should not seek coextension. Rather, we should seek a precise concept that extensionally coincides 'to a greater or lesser degree' with the common concept, and which is useful for certain technical purposes.

The three issues discussed by Tarski are the three components that an account of logical consequence will ideally have:

(1) An intuitive conception of logical consequence in natural language.
(2) A precise definition of logical consequence that applies to a formal language.
(3) A justification of why the precise definition captures the intuitive conception.

Tarski's own account is:

(1′) In natural language, logical consequence is a matter of form.
(2′) In formal language, a model-theoretic approach is correct.[4]
(3′) The definition in (2′) is an *explication* of that in (1′).

---

[3] Etchemendy, for example, writes that

arguments declared valid display the distinctively modal feature invariably attributed to such arguments ... Tarski himself, not surprisingly, recognized this guarantee to be the central feature of the 'ordinary concept'. (Etchemendy 1990, pp. 82–3)

This attributes to Tarski a commitment to a modal conception of consequence. In Etchemendy (1983), he argues against the formality of logic and must believe that Tarski would agree. Against this, Smiley (1998, §5) writes, 'in keeping with the positivism of his day, Tarski wanted nothing to do with modality'. It is unclear whether Smiley is right, since Tarski did not entirely share the positivists' aversion to traditional philosophical topics, the most telling divergence being over truth and semantic notions more generally. But if Smiley is right, that would give us further reason to think that Tarski would not have been concerned with the fallacy attributed to him by Etchemendy, which is modal. Gila Sher suggests instead that Tarski's pretheoretic notion is a blend: it 'involves two intuitive ideas: the idea that *logical consequence is necessary* and the idea that *logical consequence is formal*.' (1996, p. 654).

[4] There is some debate about whether Tarski's 1936 approach is identical to the now-standard model-theoretic approach. The debate focuses on whether Tarski's models all have the same domain or whether the domain is allowed to vary, as it is in modern model-theoretic definitions. See Mancosu (2010) for a helpful summary of the debate.

Today, a related and standard account of consequence would be:

(1″)   A natural-language sentence Φ is a logical consequence of a set of natural-language sentences Γ just if the truth of all members of Γ necessitates the truth of Φ and this holds in virtue of logical form.

(2″)   A formal sentence φ is a logical consequence of a set of formal sentences γ just if every model of γ is a model of φ (where models are understood in the usual post-Tarskian way).[5]

(3″)   Under a mapping of natural to formal language (a formalization), the concept of model-theoretic consequence in (2″) is *isomorphic* with the concept of consequence in (1″).

This account of consequence is roughly that of Stewart Shapiro (1998, 2005), which we will discuss in much greater detail in Chapter 3.

There is perhaps not as much discussion of the first component (the intuitive characterization of consequence) as there could be. Many philosophers attempt to either support or criticize some precise account of logical consequence by showing that it captures or fails to capture 'the intuitive concept of logical consequence'. This is the strategy taken by John Etchemendy (1990), and also by Beall and Restall (2006), who write that there is a 'settled core' of intuitive features of logical consequence that we should be out to capture. Field (2008, pp. 47–8) defends model-theoretic consequence on the basis that Kreisel's squeezing argument 'guarantees that intuitive validity extensionally coincides both with the technical model-theoretic sense and with derivability'.

There is a mistake here from the start, however, as Timothy Smiley points out:

> The idea of one proposition's following from others—of their implying it—is central to argument. It is, however, an idea that comes with a history attached to it, and those who blithely appeal to an 'intuitive' or 'pre-theoretic' idea of consequence are likely to have got hold of just one strand in a string of diverse theories.   (Smiley 1998, §1)

Smiley goes on to divide intuitive conceptions of logical consequence by whether they are *formal* and whether they are *necessary*. We will follow this approach. First, however, it will be helpful to consider the truth-preserving nature of logical consequence.

---

[5] We'll discuss the relationship between Tarskian and model-theoretic consequence later in this chapter.

1.3 Philonian consequence

Almost everyone agrees that truth-preservation is at least *necessary* for logical consequence, although Field (2008, p. 269) and Beall (2009) are dissenting voices.[6] Some also think that truth-preservation is *sufficient*. For example, Philo of Megara, who flourished in the fourth century BC, took an argument to be valid just if it is not the case that the premises are true and the conclusion false. For example, 'It is light' is a Philonian consequence of 'It is day' (see Gould 1970, pp. 78–82 and Kneale and Kneale 1962, pp. 128–38). Thus:

Philo Φ is a Philonian consequence of Γ iff it is not the case that all members of Γ are true and Φ false.

Philonian consequence hasn't received many supporters, but (at least one time slice of) Bertrand Russell took it seriously, and developed it to include the truth of the antecedent and practicality:

In order that it may be *valid* to infer q from p, it is necessary only that p should be true and that the proposition 'not-p or q' should be true...what is required further is only required for the practical feasibility of the inference.

(Russell 1919, p. 153)

In this quote, there is nothing more than material truth-preservation to logical consequence.[7] Likewise, Arif Ahmed (2012) has expressed some sympathy with the view. Ian Rumfitt (2015) rejects Philonian consequence because it cannot make sense of reasoning from assumptions. We can imagine someone arguing 'suppose a (resultant) force were acting on the body; *then* that body would be accelerating'. But our supposition may well be untrue, for example if no force is acting on the body in question, then anything at all will be a Philonian consequence of 'a force is acting on the body', including 'the Moon is made of cheese'.

In short, Philonian consequence *may* be plausible if we are only interested in drawing inferences from premises we know already but it looks implausible about reasoning from assumptions. What, then, should be added to Philonian consequence? By far the two most popular candidates are *necessity* and *formality*. Briefly, the difference in the two approaches is as follows.

When we teach elementary logic, we usually include a modal component in logical consequence: an argument is valid iff, if the premises are true, the

---

[6] Beall's denial that logical consequence is truth-preserving is in tension with Beall and Restall's pluralism, as we'll see in Chapter 3.

[7] For more on Russell's generally deflationist attitude to modality, see Russell (1905, p. 511).

conclusion *must* be true. On this view, 'the ball is coloured' is a consequence of 'the ball is red' since, if the second is true, the first must be true.

The proponent of a formal notion of logical consequence is likely to deny that 'the ball is coloured' is a logical consequence of 'the ball is red' because, although it is plausibly necessarily truth-preserving, it lacks a *logical* form all of whose instances are valid. For example, the corresponding argument has as one of its forms '*a* is red; *so a* is coloured', all of whose instances are necessarily truth-preserving, but the predicates '*x* is red' and '*x* is coloured' are not *logical* predicates. And it is not at all clear what other *logical* form this argument might have in virtue of which it is logical. It could be treated as enthymematic, needing a further premise such as 'all red things are coloured'. Then it would of course be logically valid, but every invalid argument is just one premise away from validity; in the extreme, we can add a contradictory premise or the conclusion as a premise. The problem for such enthymematic projects is to specify what hidden premises are permissible in order to avoid vacuity.

John MacFarlane uses the label *logical hylomorphism* for 'the tradition of characterizing logic as distinctively *formal*' (2000, p. 6). He cites Kant, Frege, and Tarski as paradigm cases (2000, pp. 20–2). Under a formal approach, expressions are classified as either logical or non-logical, with the logical expressions interpreted in a fixed way and the non-logical expressions' interpretation allowed to vary (uniformly). For example, 'Fido is a dog and Felix is a cat' logically entails 'Fido is a dog' if we treat 'and' as a logical constant, since however we interpret the premise's conjuncts (and thus the conclusion), it will contain the conclusion as a conjunct, so that the re-interpreted conclusion is true if the re-interpreted premise is. Observe that whether a formally valid consequence is necessarily truth-preserving depends on which expressions are fixed and which are variable. If for instance, 'is a woman' and 'is mortal' are both fixed, then 'Hypatia is mortal' will be a formal consequence of 'Hypatia is a woman', on account of the contingent fact that anyone who is a woman is also mortal.[8]

That point conceded, we observe that *if* we choose the logical expressions appropriately, formal consequence will be a proper subset of modal consequence: all formal consequences will also be modal consequences, but not *vice versa*. In other words, if an inference is formally valid—meaning, at least, that it has a form all of whose instances are valid—then it will also necessarily be truth-preserving. For example, 'Socrates is a person, All people are mortal; *so* Socrates is mortal' is formally valid but its truth preservation also could not be otherwise. But 'The ball is red; *so* The ball is coloured', whilst necessarily truth-preserving, doesn't have a logical form all of whose instances are valid. We return to how and why an

---

[8] If you think human beings are necessarily mortal, change the example to suit; similarly if you believe some future women will be immortal. Chapter 4 of Etchemendy (1990) has more on the dependence of various features of formal consequence on the set of expressions held fixed.

appropriate formal notion of consequence delivers the necessity of consequence in Chapter 6.

Let's now take a closer look at necessity.

1.4 The necessity of logic

The thought that logic has a distinctively *modal* character has been present since at least Aristotle:

> A syllogism is a discourse in which, certain things having been stated, something other than those things stated *results of necessity* because those things are so.
>   (Aristotle, *Prior Analytics* I.1 24b 19–21; our translation and italics)

Indeed, Rumfitt (2015, §3.2) labels the view that logical consequence has this modal character *Aristotle's Thesis*. The thesis is articulated by Timothy Smiley:

> The ingredient of necessity here (unlike formality) features in Aristotle's definition of a syllogism. It is required by his demand that proofs should produce 'understanding'... coupled with his claim that understanding something involves seeing that it cannot be otherwise. Hence, a proof needs to... [proceed] by steps that preserve necessity as well as truth.   (Smiley 1998, §1)

Incidentally, some later interpreters of Aristotle have read the clause 'because those things are so' in his definition (and its clarification in the next sentence) as laying down a requirement of relevance.[9] This would, for example rule out the validity of arguments with jointly inconsistent premises and unrelated conclusion such as 'Cats are mammals, Cats are not mammals; *so* Socrates is a man'. If that's right, Aristotle's definition imposes a requirement of relevance rejected by most contemporary logicians. Calvin Normore more generally claims that '[a]ncient logics were all in some sense relevance logics' (1993, p. 448).

Another ancient example of a philosopher who endorsed the idea that logical consequence is a modal relation was Chrysippus, who held that a conclusion follows from some premises if the truth of the premises is incompatible with the falsity of the conclusion, for example 'A man is running' implies 'An animal is moving' (see Gould 1970, pp. 80–1).

It is important to distinguish two views: (i) logical consequence is the same thing as necessary consequence, in the intuitive or metaphysical sense; and (ii) necessity

---

[9] Philipp Steinkrüger provides about a dozen such examples in his article (2015, p. 1414), which is itself a further example of the genre. The most suggestive passages in Aristotle, beyond the definition just quoted, can be found in I.23, I.25, and II.17 of the *Prior Analytics*.

is one of many characteristics of logical consequence. Tarski, for example, clearly rejected (i), and the exegetical debate is whether he subscribed to (ii). In any case, a belief in Aristotle's Thesis sometimes goes hand-in-hand with a rejection of logical hylomorphism, as we will now see.

A prominent tradition in logic distinguishes formal from material validity. The materially valid arguments are intuitively valid, in the sense of necessarily truth-preserving, but not so in virtue of form, for example 'Alan is a never-married man; *so* Alan is a bachelor' and, on Kripkean assumptions, 'the cup contains water; *so* the cup contains H$_2$O'. These arguments don't appear to have a form all of whose instances are valid, at least not a *logical* form.

In medieval logic, it was common to identify logical consequence and necessary truth-preservation. Jean Buridan, for example, held that necessarily truth-preserving arguments are valid, regardless of their form. A propos of form, he wrote: 'A consequence which is acceptable in any terms is called formal, keeping the form the same' (§1.4.2 of *The Treatise on Consequences*). But he treated formal consequences as a mere subset of the necessarily truth-preserving ones, which are the ones we should aim to capture. Generally, in medieval logic, non-formal consequences are taken just as seriously as formal ones. On this purely modal, non-formal understanding, 'a man runs; *so* an animal runs' is declared valid.[10]

Many other medieval philosophers would also reject logical hylomorphism. Indeed, Peter King states that

> Mediaeval logic is also nonformal. That is, mediaeval logic deals with inferences and assertions that do not hold in virtue of their formal features as well as those that do.   (King, 2001, p. 135)

Examples of this approach to logic are Henry Hopton, Robert Fland, Richard Billingham, and Ralph Strode, all of whom recognized non-formal consequences as valid (see Boh 2001, pp. 155–6 for references), as well as Ockham and the Pseudo-Scotus in Paris (see Read 2012 for references).[11]

In modern logic, we also find examples of philosophers who take logical consequence to be necessary truth-preservation. Etchemendy (1990, pp. 81–5) holds that '[t]he most important feature of logical consequence, as we ordinarily understand it, is a modal relation that holds between implying sentence and sentence implied'. In an earlier paper, he wrote that that there is 'little reason to think that form has much to do with logic at all' (1983, p. 334). Similarly,

---

[10] Catarina Dutilh Novaes (2011), in a discussion of MacFarlane's notion of logical hylomorphism, provides evidence that Buridan was not part of this tradition.

[11] This is the history usually told about medieval logic, though see Dutilh Novaes and Uckelman (2016) for some dissenting thoughts. They argue that medieval logicians such as Simon of Faversham would reject necessity as the crucial ingredient in logical consequence. If the history here is even messier than the standard story, that is more grist to our mill.

Stephen Read (1994, p. 264) argues that 'validity is a question of the *impossibility* of true premises and false conclusion for whatever reason', adding that the 'belief that every valid argument is valid in virtue of form is a myth' (1994, p. 264). In her recent account of logic, Catarina Dutilh Novaes (2021) takes necessary truth-preservation as a condition on logical validity. Formality, in contrast, is conspicuously absent from her list of the properties this kind of validity enjoys (2021, p. 9), and indeed she has elsewhere (2012) argued that to include it would be a mistake.

Other modern logicians do not identify logical consequence and necessary truth-preservation, but nevertheless take it as one important characteristic among many. As we will see in more detail in Chapter 2, when Beall and Restall formulate their logical pluralism, they write that 'one of the oldest features determining properly logical consequence is its necessity. The truth of the premises of a valid argument *necessitates* the truth of the conclusion of that argument' (2006, p. 14) and they include necessity in their 'settled core'. Finally, Donald Davidson holds that '*x* is a grandfather; *So x* is a father' is valid and writes '"$x > y$" entails "$y < x$", but not as a matter of form' (1967, p. 125). He believes that there are consequences that are logical because necessarily truth-preserving, but not in virtue of form. His event analysis of action sentences is motivated in part by the desire to bring more arguments within the reach of first-order logic.

1.5 The formality of logic

Tarski, as we have seen, accepted logical hylomorphism. Whether he also rejected Aristotle's Thesis that logical consequence has a modal ingredient is less clear. At any rate, he was not unique in being primarily interested in a formal notion of logical consequence.

We have already seen that Philo would reject Aristotle's Thesis, since for him material truth-preservation was sufficient for consequence. Smiley (1998, §2) argues that Philo's tutor Diodorus Cronus also held that a logical consequence 'is one that neither could nor can begin with a truth and end with a falsehood'. The 'could' and 'can' mentioned here may seem to suggest necessity, but this is unlikely for three reasons. First, Diodorus understands possibility merely as whatever is or will be true. Second, Cicero reports that Philo, Diodorus, and Chrysippus all held different accounts of consequence and reading a modal condition into Diodorus would make his consequence indistinguishable from that of Chrysippus. Further, as Mates (1953, p. 45) notes, a consequence holds 'in the Diodorean sense if and only if it holds at all times in the Philonian sense'.

We saw earlier that Aristotle's Thesis seems to define logical consequence in modal terms. But that depends on how it is read. Bernard Bolzano, one of the most interesting figures in the history of writing about logical consequence, glosses it as follows:

[s]ince there can be no doubt that Aristotle assumed that the relation of deducibility can also hold between false propositions, the 'follows of necessity' can hardly be interpreted in any other way than this: that the conclusion becomes true *whenever* the premises are true.   (Bolzano 1837, §155, fn. 1)

Importantly, Bolzano's 'whenever' just means for him that every argument of the same form is truth-preserving. It is clear from Bolzano's discussion of 'follows of necessity' that nothing distinctively modal is intended by his 'whenever'. This is compatible with necessity being a characteristic of consequence, perhaps because it falls out of a non-modal characterization; but it conflicts with the idea that the notion of consequence is modally defined. More generally, Calvin Normore maintains that 'the modal criterion for valid argument was not employed widely, if at all, in the ancient world'; and, following C.J. Martin, he dates its popularity as an 'explicitly conceived account of validity' to the twelfth-century debate between Peter Abelard and Alberic of Paris (Normore 1993, p. 448). Even then, its popularity waxed and waned over the centuries, so that for example Descartes did not insist on a modal connection between premises and conclusion, focusing instead on certainty (Normore 1993, p. 449).

Bolzano himself, writing in the first half of the nineteenth century, denies that necessity should play a role in the characterization of logical consequence. He writes that 'universally satisfiable propositions could also be said to be true by virtue of their kind or form' (1837, §147). He took the entities capable of being universally satisfiable as extralinguistic propositions, which are composed of *ideas*. Some of these ideas are variable, and others are not (a division analogous to that between logical and non-logical constants). A proposition is universally satisfiable, then, just if every replacement of its variable parts yields a true proposition. Bolzano's corresponding notion of logical consequence is:

> propositions $M, N, O, \ldots$ are deducible from propositions $A, B, C, D, \ldots$ with respect to variable parts $i, j, \ldots$, if every class of ideas whose substitution $i, j, \ldots$, makes all of $A, B, C, D, \ldots$ true, also makes all of $M, N, O, \ldots$ true'
> (Bolzano 1837, §155)

Here, again, an argument is composed of some *propositions* as premises and other *propositions* as conclusions. So, unlike model-theoretic consequence, which is usually presented as holding between a *set* of propositions (or sentences) and single proposition (or sentence), Bolzano's definition is expressed in the plural idiom. It might also appear to be multiple-conclusion but this is largely illusory: most modern treatments of multiple-conclusion systems take the conclusions disjunctively, whereas Bolzano's definition is conjunctive ('all of').

Bolzano, like Tarski, has had modal conceptions of consequence erroneously read into him. But, as Kneale (1961, pp. 96–7) argues, neither was particularly interested in modal conceptions. The only evidence ever cited to support the claim that Tarski accepted a modal notion is his statement that 'it can be proved, on the basis of this definition, that every consequence of true sentences *must* be true' (1936: 417, our italics). But Tarski's 'must' is most likely Bolzano's 'whenever': an indication that logical consequence is a generalization over forms.

In any case, Tarski and Bolzano both define logical consequence using the notion of logical form. If necessity tumbles out of their accounts, then fine; but it is not built into them. As we have noted, it is plausible that all formal consequences are also modal ones, but there are modal consequences that are not formal and these would be passed over by Tarski and Bolzano. More recently, Quine (1980) defines logical consequence in terms of logical form:

> First we define a grammatical form as *logically valid* if all sentences of that form are true. Next we define a sentence as *logically true* if it has a logically valid grammatical form. Finally we say that one sentence *logically implies* another if the conditional sentence, formed of these sentences in that order by applying 'if' to the one and 'then' to the other, is logically true.   (Quine 1980, p. 17)[12]

Quine's notions of logical truth and implication apply, unsurprisingly, to sentences, rather than Bolzano's propositions, and his arguments are single- rather than multiple-conclusion. His approach to validity is to take the conjunction of the premises as a single sentence and form the argument's corresponding conditional. The argument is valid just if this conditional is logically true, where logical truth is defined in terms of logical form.

This definition, like Tarski's first attempt (F), will fail in languages that are expressively weak. In a language that contains only the predicate F and name a, for example, Fa will be a Quinean logical truth if true at all. Quine's approach, like all substitutional ones, is highly language-dependent. Further, Quine's definition cannot account for arguments with infinitely many premises, as their conjunction cannot be formed on his account.[13] In any case, these details aside, the main point is that Quine is an example of a twentieth-century logician who defined logical consequence formally but rejected its necessary truth-preservation. For as is well-known, Quine had no truck with modality.[14]

We shall return to the formality of logic—the strand that has borne most theoretical fruit—in more detail below. We shall introduce the isomorphism-invariance account in Part II and see that it is a promising way of spelling out

---

[12] Further evidence of Quine's formal approach to logical consequence can be found in his (1952, pp. 2–3, p. 46).

[13] As e.g. George Boolos (1975) points out.

[14] See e.g. Quine (1953).

formality. And as we shall observe there, there is plenty of disagreement even within the formal camp about how this notion should be understood.

1.6 'The intuitive concept of consequence'

We have seen that it is too quick to appeal to a single, well-behaved 'intuitive concept of consequence' backed by the historical tradition. Most, but not all, take Philonian consequence to be *necessary* for logical consequence. Most, but not all, take Philonian consequence to be *insufficient* for logical consequence. Beyond that, there is no general agreement to speak of. Many authors take logic to be formal, many don't; many define logical consequence as necessary truth-preservation, many don't; many take necessity to be a feature of logical consequence, several don't; many take logical consequence as relevant, many don't. So Smiley (1998, §1) is correct that 'those who blithely appeal to an 'intuitive' or 'pre-theoretic' idea of consequence are likely to have got hold of just one strand in a string of diverse theories'.

There is thus little historical agreement on 'the intuitive concept of logical consequence'. For every philosopher or logician defining logic in formal terms, there is one denying it; and likewise for necessity. Beyond a few words about relevance, we have mostly limited our attention to formality and necessity, the two most prominent features of logical consequence. Of course, if we add other potential criteria such as suitability for modelling natural language, suitability for modelling mathematical language, axiomatizability, ontological commitment, *a priori* knowability, or others still, the picture will only become more complicated. And the interpretation of the various criteria that philosophers have read into the concept of consequence also differs between them.[15] There is no single, well-behaved intuitive concept of consequence.

How should we react to this situation? One option is logical pluralism: different precisifications are available and there is no good sense in calling any one of them *correct*. In Chapter 3, we will argue against logical pluralism. Rather than respond with pluralism, we will pick one strand of the intuitive tangle and argue that it can be well *explicated*.

Tarski, we saw, points out that 'we must reconcile ourselves from the start to the fact that every precise definition of this concept will show arbitrary features to a greater or less degree' (1936, p. 409). This language does not fit well with attempts at *coextension*, because it admits that our target is too vague and fluctuating in usage to be precisely captured. Rather, we will seek a Carnapian *explication* of the common concept, which is a precise refinement replacing a vague, intuitive

---

[15] As Warmbröd (1999, p. 513) observes, logicians' views on say, necessity and the *a priori*, vary greatly, with some even claiming that there are no necessary truths or no *a priori* truths.

conception, narrowed down for certain technical purposes (see the Prologue). When we have given up on the hope of establishing coextension, explication is an attractive alternative. It involves taking an intuitive notion and replacing it for certain purposes with a precise notion. Given our discussion of intuitions about consequence, we contend that it cannot be made uncontroversially precise.

The strand that we will pursue is *formality*. This is because, as we noted at the end of §1.3, properly explicated, formal consequence turns out to be a proper subset of modal consequence. Therefore, if we succeed in explicating *formal* consequence, the arguments we judge to be valid will also be necessarily truth-preserving. The converse doesn't hold: 'the ball is red; *so* the ball is coloured' is necessarily truth-preserving but not formally so. We will return to the question of quite *why* formality guarantees necessity at the end of Chapter 6, when we've established a precise notion of formality.

So we believe that formality is the intuition behind consequence best worth pursuing and develop it in Part II. Overall, then, we are offering an explication of the notion of formal consequence. Before we do so, however, we must argue against pluralism.

2

What Is Monism?

We have seen that there is no single intuitive conception of logical consequence; rather, there is a tangle of platitudinous features such as truth-preservation, necessity, formality and even relevance. We have also seen that logic's formality can be understood in different ways. In Part II, we will argue that the one true logic succeeds in explicating a formal conception of logic, understood in terms of topic-neutrality. Arguing for one true logic of course commits us to being logical *monists*. The present chapter will explore quite what monism amounts to, and will set up Chapter 3's discussion of pluralism.

Where do logicians stand on one versus many logics?[1] Prior to the late nineteenth century, nobody was a logical pluralist. And contemporary logicians tend to behave as monists, whatever they profess in their more philosophical moments. For although they investigate many logics, each with its own consequence relation, they usually reason about them using a single logic.[2] In fact, logicians' metatheory is typically first-order Zermelo Fraenkel Choice (ZFC) set theory or similar. If it diverges from ZFC, it's almost always an extension of it or a class version of the iterative conception ZFC partly embodies such as von Neumann Bernays Gödel (NBG); and it's usually first-order and classical. Pure logicians thus investigate all sorts of languages and logics, yet the language and logic in which these investigations are undertaken is usually the same.[3] Classical logic tends to be their foundational logic, to which they resort when they want to know whether a conclusion *really* follows from some premises—and not simply in some formal system.[4] So monism is entrenched, it chimes with the practice of logicians and mathematicians, and it is even sometimes explicitly avowed; but as we shall see in the next chapter, some philosophers have recently offered resistance to it.

We have so far been relying on an intuitive understanding of monism and pluralism. Our aim in this chapter will be to sharpen and clarify these positions. We first articulate the monist's beliefs and commitments.

---

[1] Where non-logicians stand is difficult to say, partly because they tend to conflate various notions of consequence (logical, necessary, inductive, and so on).

[2] We return to the theme of metalogical pluralism in Chapter 3.

[3] This familiar point is made with particular clarity in Williamson (2014). As he puts it: 'The maxim is: be as unorthodox as you like in your object language, provided that you are rigidly orthodox in your metalanguage.' (Williamson 2014, p. 217).

[4] Naturally, the theory is often not formalized, so that it's not always clear which principles exactly it relies on.

One True Logic: A Monist Manifesto. Owen Griffiths and A.C. Paseau, Oxford University Press.
© Owen Griffiths and A.C. Paseau 2022. DOI: 10.1093/oso/9780198829713.003.0002

2.1 Defining monism—a first pass

Monists and pluralists can agree that there are many mathematically interesting pure logics. Their beef is about whether one of these is *the* correct logic. *Logical Monists* say that one of them is, *Logical Pluralists* deny this. What, more precisely, are they disagreeing about?

2.1.1 A scheme

We begin with an approximate formulation that we sharpen over the next few pages. For present purposes, we take a logic to be a set of sentences equipped with a semantics from which a consequence relation on the logic's sentences may be defined. Call the monist's preferred logic $\mathcal{L}$. Let a formalization of an E_c^+-statement s in $\mathcal{L}$ be $Form_{\mathcal{L}}(s)$ and the formalization of a set of statements S be $Form_{\mathcal{L}}(S)$, where $Form_{\mathcal{L}}(S) = \{Form_{\mathcal{L}}(s) : s \in S\}$.[5] Take for instance the sentence 'Alex is clever and Duncan is clever and Oliver is clever'. Its formalization into first-order logic (FOL) is given by

$$Form_{\mathsf{FOL}}(\text{Alex is clever and Duncan is clever and Oliver is clever}) = Ca \wedge Cd \wedge Co$$

Most versions of logical monism are then captured by the following formulation:

Logical Monism
For all statements s of E_c^+ and sets of statements S of E_c^+, S logically entails that s iff $Form_{\mathcal{L}}(S) \vDash_{\mathcal{L}} Form_{\mathcal{L}}(s)$.

$Form_{\mathcal{L}}(S) \vDash_{\mathcal{L}} Form_{\mathcal{L}}(s)$ means: for all $\mathcal{L}$-interpretations I, if $Form_{\mathcal{L}}(S)$ is true in I then $Form_{\mathcal{L}}(s)$ is true in I.[6]

A particularly simple version of monism takes $\mathcal{L}$ as PL, propositional logic (with standard syntax and semantics). Let S be the singleton whose only element is the natural-language statement

Albert is fair and Barbara is fair.

and s the sentence

---

[5] Recall from the Prologue that E_c^+ is cleaned-up and extended English.
[6] Although we believe that the model-theoretic conception of logical consequence is correct and the proof-theoretic one mistaken, we note that a proof-theoretic monist could adopt this schema too, replacing $\vDash$ by $\vdash$. In fact, most of our discussion in Part I can be similarly reformulated along proof-theoretic lines.

Albert is fair.

The PL-formalizations of 'Albert is fair and Barbara is fair' and 'Albert is fair' are respectively $p \wedge q$ and p. That is to say,

$Form_{PL}$(Albert is fair and Barbara is fair) = $p \wedge q$

and

$Form_{PL}$(Albert is fair) = p.

So the instance of the scheme of interest is:

{Albert is fair and Barbara is fair} logically entails that Albert is fair iff $p \wedge q \vDash_{PL} p$.

Everyone agrees that the left-hand side of this biconditional is true. Since its right-hand side is also true, the biconditional itself is true.

The inadequacy of PL as the one true logic, however, is revealed when we move to arguments whose validity does not turn on propositional features. Consider the argument 'Albert is fair, therefore someone is fair'. This argument is valid, though its propositional formalization is not, since $p \nvDash_{PL} r$.

We note in passing that monists are *not* committed to thinking that 'S logically entails the sentence s' is true or false for every set of natural-language statements S and every statement s. Brouwer and Heyting, for instance, thought that there was one true logic: intuitionistic logic. Yet they did not take every claim of logical consequence to be true or false.[7] Similarly, one might be a monist and think that the one true logic is fuzzy logic. Advocates of fuzzy logic embrace a continuum of truth-values, as opposed to a mere two. On those grounds, they might reject the idea that all claims of consequence are true or false—even if in practice many of them confine the fuzziness to the object language and go classical in the metalanguage. This issue will be a major theme of Chapter 3.

2.1.2 Admissible formalizations

Although logicians are expert formalizers, a clear account of formalization is hard to come by; as Graham Priest put it, '[formalization] is a skill that good logicians acquire, but no one has ever spelled out the details in general' (Priest 2006, p. 170).

---

[7] If they had, they would have taken intuitionistic first-order logical truth to be decidable. (Consider null-premise arguments of the form $\therefore \phi$, where ϕ is a first-order sentence.) Brouwerian intuitionists, however, do not accept this commitment.

Since Logical Monism makes key use of the notion of formalization, it will be worth saying a few words about some of the potential criteria.

A first criterion is *capturing implicational relations*. Some natural-language sentences stand in this relation, others don't. For example, 'Felix is a cat' logically implies 'There's a cat' but does not logically imply 'There's a dog'. A good formalization must capture these implicational facts, or at least as many as possible, or at least as many of the most important ones (for some purpose) as possible. For instance, the propositional formalization $p \therefore q$ of the argument 'Felix is a cat, therefore there's a cat' renders it invalid, as opposed to the first-order formalization $Fa \therefore \exists x Fx$. The reason the first-order formalization is usually preferred to the propositional one is because the former captures the argument's validity whereas the latter does not. The history of logic amply illustrates the value of capturing natural-language consequence. A famous example of an argument whose validity is not captured by Aristotelian logic is 'All dogs are animals, therefore all heads of dogs are heads of animals' an argument whose validity, as we now know, can be captured in first-order logic. The criterion of capturing implication relations is *global* because it takes the whole language into account. It strives to reflect as much of the natural language's implicational 'network' as possible in the formal language.[8]

Another potential criterion is *semantic proximity*. This criterion enjoins us to formalise the natural-language sentence s as a formal sentence σ if σ may be interpreted so as to be as close in meaning to s as possible.[9] On this criterion, a formalization σ_1 of the natural-language sentence s is better than another formalization σ_2 (be it in the same logic or different ones) if some interpretation of σ_1 is closer in meaning to s than any interpretation of σ_2. For example, $\exists x Fx$ is a better formalization of 'Someone is French' than $\forall x Fx$ is, since some interpretation is closer in meaning to 'Someone is French' than any interpretation of $\forall x Fx$. The criterion therefore presupposes the existence of comparative similarity facts among propositions.[10] It is a sentential criterion because it proceeds sentence by sentence.

---

[8] Paseau (2019a) considers whether we might be able to measure how well a formalization captures implicational structure. Paseau (2019b) examines how we might distinguish between logics that do equally well with respect to this criterion, and Paseau (2021b) focuses on propositional logic's implicational limitations. See either of the first two articles for quotations from philosophers who stress the importance of preserving implicational structure.

[9] As Benson Mates writes:

> ...to formulate precise and workable rules for symbolizing sentences of the natural language is a hopeless task. In the more complicated cases, at least we are reduced to giving the empty-sounding advice: ask yourself what the natural language sentence means, and then try to find a sentence of [the formal language] $\mathfrak{L}$ which, relative to the given interpretation, has as nearly as possible the same meaning.   (Mates 1972, p. 84)

Another notable articulation of the criterion of semantic proximity may be found in Sainsbury (2001, pp. 52, 372).

[10] Paseau (2020) proposes and examines an account of facts of the form 'p_1 is more similar to p_2 than p_3 is to p_4' for propositions p_1, p_2, p_3, and p_4. Of course, the semantic proximity criterion is automatically met if σ_1 is an atomic sentence—since one can interpret it by stipulating that it is to

We have encountered two potential criteria so far, the global one of capturing implicational relations and the sentential one of semantic proximity. A potential sub-sentential criterion is that of *respecting grammatical form*. It states that a formalization should respect the grammatical form of a natural-language sentence as much as possible. In other words, the formalization's syntax should reflect the original sentence's syntax as much as possible.

But what is a sentence's syntax? Care must be taken not to impose a parochial view born by familiarity with certain languages, or families of languages. Without good reason, we should not, for instance, privilege the grammar of English over other languages, or the grammar of linear-alphabet-deploying languages (e.g. languages written using the Roman alphabet) over that of others (e.g. languages that contain ideograms, such as Egyptian hieroglyphs or Chinese characters). But even when the focus is on a single language, such as English, theorists still diverge. As an illustration, linguists have investigated extensions of first-order logic with the operator ι, which roughly stands for 'the'; more precisely, $\iota x F x$ is a term that means 'the thing that is F'. On this approach, 'The King of France is bald' should be formalized as $B(\iota x K x)$. In contrast, on Russell's classic account of definite descriptions, the same sentence should be formalized as $\exists x \forall y ((K y \leftrightarrow y = x) \wedge B x)$. The main difference between the two approaches is syntactic, as their truth conditions are identical (given the usual semantics of the ι operator and that of first-order logic). So which formalization does a better job of respecting the grammar of the sentence 'The King of France is bald'? Fans of the first approach argue that it cleaves to the structure of the English sentence more closely. Fans of the second retort that a more theoretical account of syntax is required, and that their formalization then captures the sentence's grammatical form best.

As this example illustrates and as we will see in more detail in §11.4, it would be naive to suppose that there is some ready-made, theory-neutral, notion of grammatical form one can appeal to in applying the grammatical criterion. Much work at the intersection of the philosophy of language and linguistics has gone into making sense of this and related questions. There is no need for us to enter the fray, so we rest content with a relatively uncontroversial example of how the grammatical-form criterion cuts down the available options. Take the simple sentence 'The cat sat on the mat', whose propositional formalization we can all agree should be the sentence letter p. Compare this sentence-formalization pair to the following ones:

'The cat sat on the mat and the cat sat on the mat', formalized as $p \wedge p$;

---

mean exactly what s means—whenever no interpretation of σ_2 means exactly what s means. So the semantic criterion cannot be applied to cases of this kind, in which it is trivially satisfied.

'The cat sat on the mat, and it's raining or it's not raining', formalized as $p \wedge (q \vee \neg q)$;

'The cat sat on the mat, and it's not the case that it's both raining and not raining', formalized as $p \wedge \neg(q \wedge \neg q)$.

The four English sentences (the original 'The cat sat on the mat' and the three above) are logically equivalent to one another, as are the four respective formalizations. Why then is the first sentence formalized as p, the second as $p \wedge p$, the third as $p \wedge (q \vee \neg q)$ and the fourth as $p \wedge \neg(q \wedge \neg q)$? An obvious answer is that the different formalizations respect the sentences' different grammatical forms. For instance, the first sentence is atomic, whereas the second is a conjunction of the same atomic sentence with itself. Despite their logical equivalence, each of the four sentences is formalized in a different way, to reflect its grammatical structure. Natural-language negation, conjunction or disjunction are represented in the respective formalizations by their formal-language counterparts. That much is obvious, and appreciated by the greenest of logicians. Observe in passing that p is just as good a formalization of, say, 'The cat sat on the mat and the cat sat on the mat' as $p \wedge p$ as far as our two earlier criteria are concerned. The sentence letter p is logically equivalent to $p \wedge p$, so the implicational-role criterion cannot drive a wedge between them; nor can the semantic-proximity criterion, since p can be interpreted as 'The cat sat on the mat and the cat sat on the mat'. Of the three criteria, only the grammatical one can justify a preference for formalizing 'The cat sat on the mat and the cat sat on the mat' as $p \wedge p$ rather than p.

Similarly simple examples abound. You would be doing your students a disservice if, as a logic teacher, you let them get away with first-order formalizing

It is not the case that everything is not round

as $\exists x Rx$ instead of $\neg\forall x \neg Rx$. The two predicate formalizations might be logically equivalent; but only the second is right. (The first is the correct predicate formalization of 'Something is round' instead.) In fact, one could argue that the grammatical criterion is always operative whenever one (or a few) from a large class of possible logically equivalents formalizations is adopted (such as p instead of $p \wedge p$, etc.). The moral: the grammatical-form criterion seems to do important work, even if its exact import is contested.

Since logical monism depends on the notion of formalization, we thought it worthwhile to discuss formalization a little. We mentioned three potential criteria:[11] respecting implicational structure, a criterion which operates at the level of the whole language; semantic proximity, at the sentential level; and a sub-

---

[11] Not intended as exhaustive, or entirely exclusive. Plausibly, for example, once meanings are fixed, so are implicational relations.

sentential criterion based on grammar. But our defence of monism in this book does not rest on a particular theoretical analysis of formalization, which is why we called these *potential* criteria. Like Justice Stewart with regard to pornography, the attitude we take to admissible formalization is that we know it when we see it. To call a formalization *admissible* is to say that it meets the criteria of correct formalization, whatever exactly they may be.

2.1.3 Admissible logics

Formalization is also sensitive to a choice of logic and, just as we can discuss the admissibility of formalization, we can discuss the admissibility of a logic. As we saw in Chapter 1, there are many criteria we could include in our admissible logics, such as truth-preservation, necessity, and formality. We won't list all the criteria here as our aims in this book don't depend upon doing so, but shall simply note them as we go along. As discussed in Part II, any candidate for the one true logic should include all the logical constants, and it should provide a framework for doing mathematics, since mathematical theories must be cast in a particular logic. Any logical pluralist will likewise need to provide an account of the admissibility of a logic so that not anything goes. In Chapter 3, we will see that Beall and Restall provide a clear statement of their admissibility criteria. Stewart Shapiro is less clear but for him, as we will see, roughly any logic with a useful or interesting application will suffice.

2.1.4 A true logic

We have sketched what it is to be an admissible logic and an admissible formalization into that logic. With these definitions in hand, we revise our definition of *a true logic* thus:

> $\mathcal{L}$ is a true logic just when (i) it is admissible, and (ii) there is an admissible formalization $Form_{\mathcal{L}}$ from the statements of E_c^+ to the sentences of $\mathcal{L}$ such that, for all statements s of E_c^+ and sets of statements S of E_c^+, S logically entails s iff $Form_{\mathcal{L}}(S) \vDash_{\mathcal{L}} Form_{\mathcal{L}}(s)$.

As just noted, if there is one such admissible formalization function there are likely to be many. Despite the abundance of formalizations from E_c^+-sentences to $\mathcal{L}$-ones, however, all such formalizations must return a univocal answer to whether S entails s, for each S and s. The argument is straightforward, for if $Form_{\mathcal{L}}^1$ and $Form_{\mathcal{L}}^2$ are two $\mathcal{L}$-formalizations satisfying the above, then if S entails s it follows that both $Form_{\mathcal{L}}^1(S) \vDash_{\mathcal{L}} Form_{\mathcal{L}}^1(s)$ and $Form_{\mathcal{L}}^2(S) \vDash_{\mathcal{L}} Form_{\mathcal{L}}^2(s)$, so that

$Form^1_{\mathcal{L}}(S) \vDash_{\mathcal{L}} Form^1_{\mathcal{L}}(s)$ iff $Form^2_{\mathcal{L}}(S) \vDash_{\mathcal{L}} Form^2_{\mathcal{L}}(s)$.

By similar reasoning, this biconditional also holds when S does not entail that s.[12]

2.1.5 *The* one true logic

We have defined what it is to be *a* true logic. Informally, $\mathcal{L}$ is a true logic just when it meets some admissibility conditions on a logic, and formalizations into it, also meeting some admissibility conditions, capture the natural-language consequence relation. Even more informally, $\mathcal{L}$ is a true logic just when it's the right sort of consequence-respecting mathematical model.

Our definition of a true logic, however, does not determine it uniquely. Consider for example a monist who takes first-order logic (FOL) to be the one true logic.[13] Any logic $\mathcal{L}$ that extends FOL is also a true logic, so long as admissible FOL-formalizations are all admissible $\mathcal{L}$-formalizations as well.[14] In particular, second-order logic (SOL) is also a true logic if FOL is. For if S is a set of natural-language sentences and s a natural-language sentence,

$Form_{FOL}(S) \vDash_{FOL} Form_{FOL}(s)$ iff $Form_{SOL}(S) \vDash_{SOL} Form_{SOL}(s)$

whenever the two formalization functions are equal, that is, whenever $Form_{SOL} = Form_{FOL}$.

As this example suggests, we can extract a definition of *the* one true logic from that of *a* true logic by supplementing the latter with a minimality condition. Intuitively, the one true logic is the 'least' true logic. Our framework for logical monism is developed chiefly for its own sake, and we will rest on it only lightly in the rest of the book. Accordingly, we relegate the details of how to define the 'least' true logic to an appendix.

2.2 Epistemics

We have defined what it is to be the one true logic (OTL). But how do we determine which logic OTL is and indeed whether there is just one such? These are epistemological questions, distinct from the definitional question. We answer them with an account of the applied logician's task, as we see it.

---

[12] Similarly for the proof-theoretic version of monism, which replaces $\vDash_{\mathcal{L}}$ with $\vdash_{\mathcal{L}}$.
[13] Here and elsewhere it's understood that FOL is equipped with its standard consequence relation; ditto for other common logics.
[14] By 'extends' we mean 'properly extends', unless otherwise noted.

A convenient starting point is to consider how Bertrand Russell, who put the term 'logical form' on the philosophical map, understood the applied logician's aims. He believed that each sentence has exactly one logical form, that this form is implicitly known to all competent speakers of the language, and that the applied logician's task is to discover it. As he wrote:

> In order to understand a sentence, it is necessary to have knowledge both of the constituents and of the particular instance of the form. It is in this way that a sentence conveys information, since it tells us that certain known objects are related according to a certain known form. Thus some kind of knowledge of logical forms, though with most people it is not explicit, is involved in all understanding of discourse. It is the business of philosophical logic to extract this knowledge from its concrete integuments, and to render it explicit and pure.
>
> (Russell 1914, p. 35)

Our approach in this book is entirely different from Russell's. We do not take an account of logical consequence to be epistemically constrained. As explained in the Prologue, our account is of the implication relation. This relation may or may not hold between some premises and a conclusion, irrespective of what 'most people' believe or know, implicitly or explicitly. Such an account is a scientific theory in the broad sense, just as Newtonian mechanics is a theory of planetary motion (alongside other phenomena). Naturally, the correct account of logical consequence has to be informed by reflecting on various arguments' validity or invalidity as the case may be, just as the correct account of planetary motion must be informed by experimental data about planets' trajectories.

A monist, then, believes that there is a unique best theory of the logical-consequence relation. Such a theory is broadly scientific. To put it another way, recall that at the core of logical monism is the scheme:[15]

S logically entails s iff $Form_{\mathcal{L}}(S) \vDash Form_{\mathcal{L}}(s)$.

Some philosophers might suppose that the right-hand side provides a meaning analysis of the left-hand side; it reveals what logical entailment means. To understand the left-hand-side, one must grasp its equivalence with the right-hand side. As we see it, the schema's right-hand side does not say the same thing as its left-hand side. Nor does it constitute the left-hand side; Russell's imagery of integument and extraction is inapposite. What the right-hand side can do, instead, is explain natural-language facts (i.e. the scheme's left-hand side).[16] If, say, the one true logic

---

[15] A proof-theoretic version would replace $\vDash$ with '$\vdash$'.
[16] For discussion of the notion of *explanation* involved here, see Payette and Wyatt (2018).

were propositional logic, the formal entailment $p \wedge q \vDash_{PL} p$ would explain why Albert and Barbara's being fair logically entails Albert's being fair.

What status you accord the monistic scheme may well influence your assessment of monism and pluralism's respective merits. The connections, however, are not straightforward. Shapiro proposes taking logic as a *model* of natural language, and comments that

> ...once it [the perspective of logic as a model] is adopted, the further step to pluralism is to be expected, given the nature of mathematical modeling more generally. Given the complex array of goals in modeling just about anything, there are likely to be different models that score approximately equally well on the various criteria, taken together.   (Shapiro 2014, p. 65; see also p. 49)

To what extent our account of what we are up to in this book matches Shapiro's 'logic-as-model' perspective is an interesting question. Irrespective of the answer, we beg to differ on one point. The logic-as-model perspective is certainly compatible with pluralism; but it need not inexorably lead to it. That the General Theory of Relativity is a model of gravitation does not *eo ipso* disqualify it from being *the* single correct comprehensive theoretical account of gravitation. Similarly for any scientific model. That a model ignores some properties the better to concentrate on others does *not* preclude its being uniquely best. In giving arguments, we do lots of things. One aspect of what we sometimes do is give, or try to give, valid arguments. Perhaps there is a best model of this practice, perhaps there isn't. The logic-as-model perspective is compatible with both monism and pluralism.

We have characterized the aim of our approach as that of giving a scientific theory (in the broad sense) of natural language's logical-consequence relation(s). Although we have not fussed over the difference (and will continue not to), we generally favour this description to that of giving a model. In science, the term 'model' tends to be used in a more specific sense than the word 'theory' does. For example, when a research team advertises for a 'model-builder', they are looking for something more specific than a theory-builder. Typically, a model involves various simplifications and idealizations, often gross ones. When Shapiro is read as using the word 'model' in that sense, his point looks better. Be that as it may, we prefer to characterize our account as a 'theory' rather than a 'model', and see no general reason to suppose from the outset that it cannot be uniquely correct.

Applied logic thus offers a theory of the natural-language consequence relation. To evaluate which logic is the one true logic, and indeed whether there is one, one must resort to the usual criteria of theory choice. What exactly these are is controversial, and how exactly they relate to one another even more so. In particular, it is unclear exactly what should count as the data, for example the judgements of the folk, or of logicians. Nonetheless, we can cite as relevant criteria:

extensional adequacy, that is capturing the data of natural-language entailment, the closer the better; simplicity; strength; unity; and so on. To list these criteria in this way is not to suggest that they should all have the same weight; clearly, some are more important than others. Furthermore, satisfaction of the criteria is not an all-or-nothing affair; it comes in degrees. This should be particularly obvious in the case of extensional adequacy: logic $\mathcal{L}_1$, for instance, may respect more arguments' validity than logic $\mathcal{L}_2$ without respecting them all.

We shall see in Chapter 3 that Beall and Restall, articulate contemporary defenders of logical pluralism, have erred by building in too few constraints on applied logic (the 'settled core' of consequence in their terminology). We also note in passing how precarious logical pluralism is. Anything other than the exact equilibrium position between the top contenders, judged by these criteria, implies that monism wins the day. In other words, pluralism requires a hard-to-achieve perfect balance; the slightest difference is enough to tip the scales. And as we point out in Chapter 3, monism itself has as good a claim to be part of the settled core of consequence as anything. As for the idea that there are different clear winners in different contexts, we will see that this isn't a form of pluralism after all.

Appendix: the least true logic

We defined *a* true logic in §2.1 and said that *the* true logic is the least one among these. This appendix spells out a promising way of analysing what 'least' here means. We have relegated this material to an appendix because it is tentative and technical.

Our suggestion is that the one true logic should *embed* into any other. More precisely, consider two logics $\mathcal{L}_1$ and $\mathcal{L}_2$, with respective sets of sentences $Sen(\mathcal{L}_1)$, $Sen(\mathcal{L}_2)$ and respective consequence relations $\vDash_{\mathcal{L}_1}, \vDash_{\mathcal{L}_2}$. The map $j : Sen(\mathcal{L}_1) \to Sen(\mathcal{L}_2)$ is a *conservative translation* just when, for all $\Gamma \subseteq Sen(\mathcal{L}_1)$ and all $\delta \in Sen(\mathcal{L}_1)$,[17]

$\Gamma \vDash_{\mathcal{L}_1} \delta$ if and only if $j(\Gamma) \vDash_{\mathcal{L}_2} j(\delta)$.

Equivalently, $j : Sen(\mathcal{L}_1) \to Sen(\mathcal{L}_2)$ is a conservative translation when, for all $\Gamma \subseteq Sen(\mathcal{L}_1)$ and $\delta \in Sen(\mathcal{L}_2)$,

(i) if $\Gamma \vDash_{\mathcal{L}_1} \delta$ then $j(\Gamma) \vDash_{\mathcal{L}_2} j(\delta)$;
(ii) if $\Gamma \nvDash_{\mathcal{L}_1} \delta$ then $j(\Gamma) \nvDash_{\mathcal{L}_2} j(\delta)$.

A *bijective conservative translation* is a conservative translation $j : Sen(\mathcal{L}_1) \to Sen(\mathcal{L}_2)$ that is also a bijection. Since two logics conceived in this way that are related by a bijective conservative translation are the same up to relabelling of sentences, we may call a bijective conservative translation a *consequence isomorphism* and two logics related in this way *consequence isomorphic*.[18] Observe in passing that if j is a consequence isomorphism then so is $j^{-1} : Sen(\mathcal{L}_2) \to Sen(\mathcal{L}_1)$. We also note that in the relevant literature (e.g. Pinheiro Fernandes 2018), 'translation' usually relates to deductive consequence; but here we apply

---

[17] Where $j(\Gamma) = \{j(\gamma) : \gamma \in \Gamma\}$.
[18] This agrees with the terminology in Paseau (2019a).

it to semantic consequence. Similarly, a *consequence embedding* is an injective conservative translation.

We may take $(Sen(E_c^+),\therefore)$ to be the ordered pair of (cleaned-up and extended) natural language and its consequence relation, and extend the above definitions to cover this language-consequence pair as well. Thus a conservative translation from English into a logic $\mathcal{L}$ is a map (formalization function) from the sentences of English to those of $\mathcal{L}$ such that a set S of English sentences implies s iff S's formalization (i.e. S's image under the map/formalization) implies s's formalization in $\mathcal{L}$. $\mathcal{L}$ is then *a* true logic just when there is an admissible conservative translation from E_c^+ to $\mathcal{L}$. (This map has domain $Sen(E_c^+)$ and codomain $Sen(\mathcal{L})$ and satisfies the conditions on formalization; as we saw earlier, if there is one such map there will typically be many.) OTL is *the* true logic if it is a true logic, and it consequence embeds into any other true logic $\mathcal{L}$, so that the composition of an admissible conservative translation from E_c^+ to OTL and the consequence embedding from OTL to $\mathcal{L}$ is identical to any given admissible conservative translation from E_c^+ to $\mathcal{L}$. For short, we may say that OTL has the *embedding property*, or that any formalization *factorizes* via OTL. Expressed in a diagram:

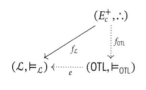

In the diagram, the solid arrow indicates a conservative translation and a dotted arrow a construction. The diagram depicts the fact that if $f_{\mathcal{L}}$ is an admissible conservative translation from E_c^+ to $\mathcal{L}$ then there is a consequence embedding e from OTL to $\mathcal{L}$ and an admissible conservative translation f_{OTL} from E_c^+ to OTL such that the diagram commutes, that is, $f_{\mathcal{L}} = e \circ f_{OTL}$.[19] For example, if FOL is the one true logic, then any admissible conservative translation from E_c^+ (and its consequence relation) into SOL decomposes in this two-step manner.

Observe that two logics may be consequence-isomorphic to one another although no admissible formalizations from E_c^+ into each of them exists to instantiate the commutative diagram above. In other words, $\mathcal{L}$'s realizing the one true logic need not imply that all its consequence-isomorphs also do so. Admissibility constraints on formalization may prevent some $\mathcal{L}$-consequence isomorphs from also realizing the one true logic, as may the logic's admissibility itself.

Our definition of *the* one true logic from that of *a* true logic requires a proviso. The definition does not preclude there being more than one such true logic, casting doubt over our use of the definite article. That use would be justified if the following condition obtains. Suppose that $\mathcal{L}_1$ and $\mathcal{L}_2$ are *both* true logics according to our definition; then (i) $\mathcal{L}_1$ and $\mathcal{L}_2$ are consequence isomorphic, and (ii) this isomorphism respects admissible formalizations (meaning that an $\mathcal{L}_1$-admissible formalization composed with this isomorphism is an $\mathcal{L}_2$-admissible formalization and vice versa). In that case, $\mathcal{L}_1$ and $\mathcal{L}_2$ may for our purposes be considered the same logic. For example, there is a clear sense in which FOL with constants a_i (where i is a natural number) and predicate letters F_{ij} (with i and j natural numbers, i indicating the predicate's adicity) is identical to FOL with constants b_i and predicate letters

---

[19] A proof-theoretic monist can give a similar account, replacing model-theoretic with proof-theoretic consequence throughout.

G_{ij}.[20] Under such circumstances, we could define the one true logic as the isomorphism type realized by any true logic $\mathcal{L}$ with the embedding property. This would justify our calling it 'the one true logic'. It would also accord well with *some* aspects of logicians' usage, as they do not typically identify a logic with a particular choice of vocabulary. They instead think of it as what its various presentations have in common, vocabulary differences aside.[21]

For all we know, it may be that any two candidates for the one true logic do satisfy the conditions in the previous paragraph, *viz.* if they consequence embed into each other then they are consequence isomorphic in an admissible-formalization-respecting way. Yet this condition does not in general obtain. Recall that sets have the Schröder-Bernstein property, *viz.* if there's an injection from set A to set B and an injection from set B to set A then there's a bijection between A and B. Logics, abstractly conceived, however, lack the analogous 'Schröder-Bernstein' property. That is to say, there are logics $\mathcal{L}_1$ and $\mathcal{L}_2$ such that $\mathcal{L}_1$ consequence embeds into $\mathcal{L}_2$ and $\mathcal{L}_2$ consequence embeds into $\mathcal{L}_1$, yet there is no consequence isomorphism between the two, still less an admissible-formalization-respecting one. A proof of this abstract possibility is given at the end of this appendix. The fact that such a situation could arise does *not* mean that it does so in the case at hand. For it may be that the candidates for the one true logic *are* all consequence-isomorphic to each other. (Compare: the fact that there is an embedding from topological space A to topological space B and an embedding from B to A does not in general imply that A and B are homeomorphic, that is, isomorphic as topological spaces; but of course some pairs of mutually embeddable topological spaces are homeomorphic.)

In the spirit of worst-case-scenario planning, suppose that the situation noted in the previous paragraph were to arise. That is to say, suppose it turns out that logics $\mathcal{L}_1$ and $\mathcal{L}_2$ both have the embedding property noted earlier (and in particular that a formalization into $\mathcal{L}_1$ factorizes via $\mathcal{L}_2$ and vice versa), yet there is no admissible-formalization-respecting consequence isomorphism between $\mathcal{L}_1$ and $\mathcal{L}_2$. The natural thing to say would be that the one true logic is the isomorphism type of the image of $(Sen(E_c^+), \therefore)$ under any admissible formalization into a least true logic. This is a single structure or isomorphism type since, by definition, it's the structure consequence-isomorphic to that of English. (This definition also applies in the case in which the least true logics all realize the same consequence-isomorphism type.) Naturally, in such a case, the best way to get a handle on the resulting structure may be to consider it as a substructure of a broader logic; understanding all of $\mathcal{L}$ may be the way to get to grips with the fragment(s) of $\mathcal{L}$ to which English is consequence-isomorphic. However that may be, we now have a definition of the one true logic that covers the—somewhat abstruse—possibility just noted.

We conclude this appendix with the promised example of two logics $\mathcal{L}_1$ and $\mathcal{L}_2$ such that the first consequence embeds into the second and the second into the first, but that are not consequence-isomorphic. Let the sentences of $\mathcal{L}_1$ and $\mathcal{L}_2$ both be the natural numbers (i.e. $0, 1, 2, \ldots$). In $\mathcal{L}_1$, $\Gamma \vDash \delta$ iff (i) Γ is an infinite set, or (ii) Γ is a non-empty finite set the sum of whose sentences is greater than or equal to δ, and δ is an *even* power of 2 $(1, 4, 16, 64, \ldots)$. In $\mathcal{L}_2$, $\Gamma \vDash \delta$ iff (i) Γ is an infinite set, or (ii) Γ is a non-empty finite set the sum of whose sentences is greater than or equal to δ, and δ is an *odd* power of 2 $(2, 8, 32, 128, \ldots)$. For example, $\{3, 7\} \vDash_{\mathcal{L}_1} 4$ because $3 + 7 = 10 > 4$ and 4 is an even power of 2, but $\{3, 7\} \nvDash_{\mathcal{L}_2} 4$ because 4 is not an odd power of 2.

---

[20] Assuming no differences other than these vocabulary ones.

[21] Although we recognize that it does not fit *all* of logicians' informal identity criteria for logics. Logicians, for example, often do not care whether the set of atomic sentence letters is countably or uncountably infinite, even though the difference affects consequence isomorphism.

We prove that (i) $\mathcal{L}_1$ consequence embeds into $\mathcal{L}_2$; (ii) $\mathcal{L}_2$ consequence embeds into $\mathcal{L}_1$; but (iii) $\mathcal{L}_1$ and $\mathcal{L}_2$ are not consequence-isomorphic. To see (i), note that the injection $n \mapsto 2n$ is a consequence embedding from $\mathcal{L}_1$ to $\mathcal{L}_2$. Similarly, for (ii), the injection $n \mapsto 2n$ is a consequence embedding from $\mathcal{L}_2$ to $\mathcal{L}_1$.

Finally, to prove (iii), suppose f is a conservative translation from $\mathcal{L}_1$ to $\mathcal{L}_2$. Call a self-implying sentence in either logic one that implies itself, that is, a sentence α such that $\alpha \models \alpha$. In $\mathcal{L}_1$, these are all and only the even powers of 2, and in $\mathcal{L}_2$ they all and only the odd powers of 2. Since the self-implying sentences of $\mathcal{L}_1$ and $\mathcal{L}_2$ are well-orderable (say that α precedes β if both are self-implying and $\beta \models \alpha$ but $\alpha \not\models \beta$), $f(1) = 2, f(4) = 8$, etc. Likewise, for either logic, 0 is the only sentence whose addition to a premise set never makes a difference to the argument's validity, so we must have $f(0) = 0$. Hence if a is the sentence (i.e. number) such that $f(a) = 1$ then $a > 1$. Now $a \models_{\mathcal{L}_1} 1$ since $a > 1$ and 1 is an even power of 2; but $f(a) \not\models_{\mathcal{L}_2} f(1)$, since $1 \not\models_{\mathcal{L}_2} 2$, because $1 < 2$.

Naturally, the addition of further admissibility conditions on logics may block such counterexamples to this Schröder-Bernstein-like property.

3

Against Pluralism

We saw in the last chapter that logical monists take there to be *one true logic*. Logical pluralism is the view that there are many, equally correct, logics. The logical pluralist seeks to undermine the debates between logical monists—such as the debate between the classical and intuitionistic logician—by showing that there is no real disagreement.

We also saw that the way to articulate pluralism is to argue that there is more than one genuine logical consequence relation. We start by thinking a little more about pluralism and then offer a taxonomy of such positions. Our focus will be on two twenty-first-century forms of pluralism to have received book-length treatment: Beall and Restall (2006) and Shapiro (2014), the latter explicitly building on the former.[1]

3.1 Extensionally divergent formalizations

We saw above that, typically, if there is one admissible formalization from E_c^+ into $\mathcal{L}$ there will be many. What if there are two equally acceptable formalizations $Form_{\mathcal{L}}^1$ and $Form_{\mathcal{L}}^2$ that do *not* result in extensionally identical accounts of logical consequence? That is, what if there is a set S of E_c^+-statements and an E_c^+-statement s such that $Form_{\mathcal{L}}^1(S) \vDash_{\mathcal{L}} Form_{\mathcal{L}}^1(s)$ but $Form_{\mathcal{L}}^2(S) \nvDash_{\mathcal{L}} Form_{\mathcal{L}}^2(s)$? $\mathcal{L}$ would then be the one true logic but admit of different, extensionally distinct, formalizations.

Whether the resulting position is a form of logical monism as generally conceived is not entirely clear. It is hard to count it as a form of logical pluralism, as it concedes that there is a unique best logic, $\mathcal{L}$. Moreover, since the analysis of logical consequence only makes reference to the single logic $\mathcal{L}$ the resulting position arguably counts as a form of monism, our definition in §2.1 notwithstanding. As we do not know of anyone who has put forward this challenge to monism, rather than waste ink on so hypothetical a challenge we pass over it to consider more live ones.

---

[1] See Roy Cook (2010), Gillian Russell (2014), and Florian Steinberger (2019) for overviews.

One True Logic: A Monist Manifesto. Owen Griffiths and A.C. Paseau, Oxford University Press.
© Owen Griffiths and A.C. Paseau 2022. DOI: 10.1093/oso/9780198829713.003.0003

3.2 Horses for courses: languages and domains

Logical monism is a claim about languages and their extensions. One might try to challenge it by arguing that different languages require different logics. A related challenge is that different parts of a single language require different logics. In fact, these two 'horses for courses' challenges are identical at root, and neither is a threat to monism.

3.2.1 Horses for courses: interlinguistic

Let us start with the interlinguistic version of the horses-for-courses challenge. Since there are several natural languages, one might conceivably maintain that logic $\mathcal{L}_1$ is the right logic for one, $\mathcal{L}_2$ the right logic for another (where $\mathcal{L}_1 \neq \mathcal{L}_2$), and so on. This standpoint is compatible with monism assuming that the logical consequence relations in different natural-language extensions don't conflict. A natural no-conflict condition would be:

Suppose that E_{1c}^+ and E_{2c}^+ are disambiguated and extended natural languages, $\mathcal{L}_1$ is the right logic (with formalization $Form_{\mathcal{L}_1}$) for E_{1c}^+ and $\mathcal{L}_2$ (with formalization $Form_{\mathcal{L}_2}$) the right logic for E_{2c}^+.[2] Suppose that S_1 is a set of E_{1c}^+-statements and s_1 is an E_{1c}^+-statement, and that the translation S_2 of S_1 in E_{2c}^+ and the translation s_2 of s_1 both exist.[3] Then $Form_{\mathcal{L}_1}(S_1) \vDash_{\mathcal{L}_1} Form_{\mathcal{L}_1}(s_1)$ iff $Form_{\mathcal{L}_2}(S_2) \vDash_{\mathcal{L}_2} Form_{\mathcal{L}_2}(s_2)$.

Suppose the no-clash condition holds for all pairs of languages E_{1c}^+ and E_{2c}^+, and all S_1, s_1, S_2, and s_2 (whenever these exist). Then any reason to use $\mathcal{L}_1$ for E_{1c}^+ and $\mathcal{L}_2$ for E_{2c}^+ is due to a difference in the languages' expressive power: using E_{1c}^+ one can advance some arguments that cannot be advanced using E_{2c}^+ or vice versa (or both).

In particular, the existence of mutually untranslatable languages with different logics does not threaten monism. Suppose for example that the meanings classical and intuitionistic logicians assign to the logical constants are distinct, resulting in two languages E_1 and E_2 respectively. Then classical logic would be correct for E_1-consequence and intuitionistic logic for E_2-consequence. In particular, there is no set of sentences S and sentence s of either of these two languages such that S entails that s according to one logic but not according to the other (given admissible formalizations).[4]

---

[2] We use the example of two languages for simplicity. The discussion generalizes in the obvious way to more than two.

[3] The translation of a set of statements is the set consisting of the translation of each of the statements.

[4] The clash between classical and intuitionist logicians has not traditionally taken this eirenic form. Many intuitionists from Brouwer to Dummett have claimed that the classical meanings are in a deep

3.2.2   Horses for courses: intralinguistic

That was the interlinguistic version. Now for the intralinguistic version of the horses-for-courses challenge. Suppose E_c^+ is an extension of English such that $\mathcal{L}_1$ (with formalization $Form_{\mathcal{L}_1}$) is the right logic for some E_c^+-arguments, whereas $\mathcal{L}_2$ (with formalization $Form_{\mathcal{L}_2}$) is the right logic for the rest. Such a supposition is, once more, entirely compatible with monism. As long as no single argument is judged to be valid by one logic and invalid by another, monism is not imperilled. The one true logic is then simply the disjunction of $\mathcal{L}_1$ and $\mathcal{L}_2$, each sovereign in their respective spheres. The only way to generate a counterexample to monism is to find an argument that's valid according to $\mathcal{L}_1$ but not $\mathcal{L}_2$, with both $\mathcal{L}_1$ and $\mathcal{L}_2$ being equally good models of natural-language consequence.

In other words, the intralinguistic version is also just a form of monism. Recall Priest's (2006, p. 196) point that an interesting form of pluralism must hold the application of logic fixed as the 'canonical' application of logic to deductive reasoning. So a *bona fide* form of pluralism would argue that there are many logics that apply equally well to the core application of deductive reasoning. We will see that this is what Beall and Restall, and Shapiro, offer.[5]

We can now see how similar the interlinguistic and intralinguistic versions of the horses-for-courses observation are. The only difference is over whether the different classes of argument requiring different logics are drawn from different languages or the same one. If in the interlinguistic case we define a broader single language by taking the union of the two, we convert this case to the intralinguistic one.

Now we take our definitions of monism and pluralism to explicate the core opposition underlying the contemporary debate, and as a corollary that of pluralism too. We do not deny that some philosophers who have claimed that different logics are required for sentences with different meanings might have thought of their project as pluralist.[6] Pluralism as here defined, however, is a stronger thesis.

It is a corollary of our definition in Chapter 2 that a 'disjunctive' logic may still be the one true one. To illustrate this, let's go back to the intralinguistic version of the challenge, in which $\mathcal{L}_1$ is the right logic for one class of arguments and $\mathcal{L}_2$ for another, where the two classes partition the class of all arguments. Taking $\mathcal{L}$ as the disjunction of $\mathcal{L}_1$ and $\mathcal{L}_2$, the logician who proposes that an argument of the first type is valid iff its $\mathcal{L}_1$-formalization is $\mathcal{L}_1$-valid and an argument of

---

sense flawed or even incoherent, so that the only meaning one may legitimately assign to natural-language sentences vindicates intuitionistic logic. Unique characterization results, going back to Popper (1948) and elaborated subsequently, show that classical and intuitionistic connectives cannot coexist in the same language.

 [5] This for us is an *application* of logic because logic is about implication rather than reasoning, as explained in the Prologue.

 [6] Batens (1985) is perhaps an example. As we will see in §3.3, Beall and Restall are keen to avoid this sort of pluralism, which they attribute to Carnap.

the second type is valid iff its $\mathcal{L}_2$-formalization is $\mathcal{L}_2$-valid counts as a monist according to our definition. After all, this account of logical consequence conforms to our characterization of monism. For example, if mathematical discourse uses the material conditional, we translate it into our logic by one symbol $\supset$, and if non-mathematical discourse uses another, we might translate it into our logic by another, say $\rightarrow$, each connective having its own semantic rules. Or if disjunction behaves differently in sentences about quantum phenomena, we might translate it differently from normal disjunction.

Another way to argue for a similar conclusion is to suppose that the logical-consequence relation depends on its relata. Equally permissible choices of relata result in extensionally different accounts of logical consequence. This idea is articulated and defended in Russell (2008).

In sum, we must distinguish the idea that there is one true logic from the additional requirement that this one true logic should in some sense be 'unified'. To reject the former is to reject monism; but to reject the latter is perfectly compatible with monism. Disjunctive monism is not the same as pluralism.

3.3 Pluralism proper

The final and most popular way to reject Logical Monism is to claim that there is more than one genuine logical consequence relation. On this view, at least two logics provide extensionally different, yet equally acceptable, accounts of natural-language consequence. This is the position defended, for example, in Beall and Restall (2006) and, in a different key, Shapiro (2014). These sorts of challenges to monism are worth taking very seriously.

3.3.1 Beall and Restall's modest pluralism

Beall and Restall (2006, pp. 78–9) are clear that the interlinguistic view is uninteresting. Instead, they keep the object language fixed and put forward a version of pluralism by varying the meaning of *metalogical* notions such as 'logical consequence', 'logical truth', and 'validity'. For example, suppose the classicist says '$P \vee \neg P$ is a logical truth' and the intuitionist '$P \vee \neg P$ is not a logical truth'.[7] They should be understood as claiming '$P \vee \neg P$ is a logical truth$_{classical}$' and '$P \vee \neg P$ is not a logical truth$_{intuitionist}$', respectively. There is then no disagreement between the parties. This metalogical position *is* a form of pluralism proper.

---

[7] Where P is an interpreted statement and $\vee$ and $\neg$ mean 'or' and 'it's not the case that' respectively. It is perhaps moot whether the sentential connectives 'or' and 'it is not the case that' are truth-functional in ordinary English. But English here includes the technical vocabulary of mathematics so these interpretations *are* available, by stipulation.

Beall and Restall note that different logical-consequence relations can be formed by instantiating the Generalized Tarski Thesis (2006, p. 29):

GTT An argument is valid$_x$ if and only if, in every case$_x$ in which all of the premises are true, so is the conclusion.

GTT is a schematic expression of the truth-preservation of validity. Different definitions of 'valid$_x$', such as classical, intuitionist, relevant, and necessarily truth-preserving, are formed by instantiating different sorts of 'case$_x$'. Throughout, the language used is that of first-order logic with the usual selection of logical constants.

Beall and Restall usefully summarize their position as follows (2006, p. 35):

1. The settled core of consequence is given in GTT.
2. An instance of GTT is obtained by a *specification of the cases$_x$* in GTT, and a specification of the relation *is true in a case*. Such a specification can be seen as a way of spelling out *truth conditions*.
3. An instance of GTT is *admissible* if it satisfies the settled core of consequence, and if its judgements about consequence are necessary, normative, and formal.
4. A *logic* is given by an admissible instance of GTT.
5. There are *at least two* different admissible instances of GTT.

In the same passage, they provide four admissible instances of GTT:

C If we interpret cases as Tarskian models (TMs), we obtain classical logic.
N If we take cases to be possible worlds, we obtain a 'necessary truth-preservation' logic that validates, e.g. 'the ball is red; *so* the ball is coloured'.
I If we take cases as constructions, we obtain an intuitionistic logic that invalidates, e.g. *double negation elimination*, because constructions can be incomplete. Tarskian models are a proper subset of constructions: they are the complete constructions.
R If we take cases as situations, in the tradition of such relevant logicians as Jon Barwise and John Perry (1983), we obtain a relevant logic that invalidates e.g. disjunctive syllogism, because situations can be inconsistent. Tarskian models are a proper subset of the situations: they are the complete and consistent situations.

We call Beall and Restall's a *modest* pluralism, since it endorses a relatively small number of logics. This will contrast with *eclectic* pluralism, to be discussed shortly.

Beall and Restall remind us that their pluralism holds 'within a language as well as between languages' (2006, p. 79). Thus they endorse logics with the same language, which nevertheless disagree about the validity of certain arguments. They contrast this with Carnap's (1934) tolerant attitude to logics, where different, equally correct logical consequence relations in different languages are endorsed. Beall and Restall's position is more radical: they endorse logics with the same language, which nevertheless disagree about the validity of certain arguments.

To connect this to our Chapter 2 account of monism, Beall and Restall can be read as agreeing that a logic $\mathcal{L}$ is true just when it is admissible (truth-preserving, necessary, formal, and normative), and there is an admissible formalization function from E_c^+ to the sentences of $\mathcal{L}$ such that, for all statements s of E_c^+ and sets of statements S of E_c^+, S logically entails that s iff $Form_\mathcal{L}(S) \vDash_\mathcal{L} Form_\mathcal{L}(s)$. They would disagree, of course, that there is a single 'least' true logic: they endorse at least necessary truth-preservation, classical, intuitionist, and relevance logics.

3.3.2 Shapiro's eclectic pluralism

Shapiro writes that 'the present study begins where Beall & Restall leave off' (2014, p. 38). Although Beall and Restall do discuss many logics as precisifications of the GTT, Shapiro believes that they also rule many important logics out.

Since classical logic is amongst the instances of GTT deemed admissible by Beall and Restall, Shapiro (2014, p. 39) notes that they must accept

(*) $\forall x(\Phi(x) \vee \neg\Phi(x))$

as a logical truth. And, since classical logic is an admissible logic, (*) must be a *necessary* truth. The negation of (*) is therefore a necessary falsehood. So any theory that contains this negation must be rejected. Intuitionistic analysis and intuitionistic arithmetic with Church's thesis are such theories, and they must be rejected on Beall and Restall's view.

Nevertheless, Shapiro maintains, these theories have a great deal of mathematical and philosophical interest, and that suffices for their being worthy of our attention. The theories ruled out by Beall and Restall motivate Shapiro to formulate a version of pluralism that is even more permissive. He calls his view *eclectic pluralism*.

Shapiro's view is explicitly a form of *relativism*—a label which Beall and Restall seem keen to avoid. He follows Crispin Wright's (2008) understanding of *folk relativism*:

> 'folk relativism' concerning a given predicate Φ is the thesis that there is no such thing as simply being Φ. If Φ is folk-relative, then, in order to get a truth-value for a sentence in the form 'a is Φ', one must indicate a value for something else, sometimes called a parameter. (Shapiro 2014, p. 182)

For example, to judge whether an object is to the left, one must first specify a perspective; to judge whether two events are simultaneous, we must specify a frame of reference; and, plausibly, to judge whether something is tasty, we must specify a palate. In the case of logical consequence, we must specify a logic, as in 'ϕ is a logical consequence of Γ in logic L'. Until we have specified parameter L, we cannot judge whether some sentence follows from some others.

To this folk relativism about logical consequence, Shapiro adds a Hilbertian view of mathematical theories. For Hilbert, any consistent axiomatization characterizes a mathematical structure. Not all structures will be interesting and worthy of study, of course, but many will be. How do we know whether some structure is interesting and worthy of our attention? We go to the mathematicians. If they are investigating a given structure, we should take that seriously.

Again, then, we can see Shapiro as accepting that a logic is true just when it is admissible and can be supplemented with an admissible formalization. But again, he would deny that there is a single 'least' true logic, since his extremely weak standard of admissibility allows many logics, many more than Beall and Restall.

In sum, pluralism proper endorses at least two extensionally different but equally correct consequence relations for natural language. We have seen that Beall and Restall's modest pluralism is of this sort, and that Shapiro's can also be understood in this way.[8] Beall and Restall's modest view endorses (at least) necessary truth-preservation, intuitionistic, relevant, and classical logics, since these satisfy the settled core of consequence. Shapiro's eclectic approach endorses many more logics, as many as have found fruitful application. This seems to be the extent of Shapiro's criteria of *admissibility*.

We will first consider some problems for Beall and Restall's pluralism, and Shapiro's pluralism considered in isolation and then use our observations to pose a general problem for any version of pluralism proper.[9] Briefly, any pluralist will want to engage in some metalogical reasoning. We will argue that the eclectic will struggle here, given the large number of logics endorsed. The obvious response is to endorse a more modest pluralism. The problem here, we argue—based on the considerations about the intuitive conception of consequence in Chapter 1—is that there is no well-motivated stopping point.

We end the section with a diagram of the situation, in terms of the numbers of logics endorsed: nihilism, monism, modest pluralism, and eclectic pluralism. (Nihilism, the view that there is no correct logic, will be introduced in §3.7.)

---

[8] Shapiro (2014, ch. 5) discusses other varieties of logical pluralism but we will focus on this stronger form, which qualifies as pluralism proper.

[9] Part of our argument is metalogical. The problems that metalogic poses for the logical pluralist are also discussed in Sereni and Sforza Fogliani (2020). The arguments they examine are distinct from ours, and are not endorsed by them.

Nihilism ——— Monism ——————— Modest ———————— Eclectic
(none)   (one)   (a few)   (many)

3.4 Beall and Restall's modest pluralism

3.4.1 The 'settled core' of consequence

In the last chapter, we saw that the settled core of consequence for Beall and Restall contains necessary truth-preservation, formality, normativity, and nothing else. What justifies this? They don't directly answer this question, but offer some hints. We find appeal to 'the tradition': 'These features are central to the tradition, and any account of logic must take account of them' (2006, p. 14). And we find appeal to 'the notion': 'We hold that the notion settles some but not all features of any candidate relation of logical consequence' (2006, p. 29).

Some features get explicitly dropped. Elsewhere, they acknowledge that a

long and rather formidable tradition claims that neither '$\exists x(x = x)$' nor '$\exists x(Fx \vee \neg Fx)$' is Really Valid; logic, in this tradition, allows for the empty case, but Tarskian-style cases are never empty.   (Beall and Restall 2000, p. 480)

But they then set it aside:

If that notion is not so settled, then the matter of existential commitment provides another avenue for plurality. For present purposes, we leave that avenue open for further debate.   (Beall and Restall 2006, p. 77).

Axiomatizability also gets considered:

Logical consequence calls for both the axiomatisability of the consequence relation, and the expressibility of full second-order consequence.
(Beall and Restall 2006, pp. 103–4)

But it gets ruled out:

Nothing in the absolutely settled core of the notion of logical consequence dictates that the relation of logical consequence be axiomatisable.
(Beall and Restall 2006, p. 77)

So, the test for settledness is empirical. We look to the tradition and the state of debates about these various features, and we see that a consensus has been reached on some (necessity, formality, and normativity), but not others (axiomatizability, lack of existential commitment). And more than one logic possesses the settled features.

In a review of Beall and Restall's book, one of us noted that 'the addition of further constraints genuinely threatens each of the four accounts' (Paseau 2007, p. 393). If lack of existential commitment were added, for example, only (some) free logics would be admissible. If axiomatizability were included, full second-order logic would fail to be admissible. If facts about logical consequence must be *a priori* knowable, necessary truth-preservation logic would fail (on the Kripkean assumption of necessary *a posteriori* truths). Moreover, if we include 'suitability for modelling mathematical discourse' in the settled core (see Chapter 7), that would presumably rule out necessary truth-preservation and relevant logics and arguably intuitionistic logic.[10]

Beall and Restall need it to be the case, therefore, that some features (those compatible with pluralism) really are settled, whilst others (incompatible with pluralism) are not. As we saw in Chapter 1, however, the supposedly settled elements are far less settled than Beall and Restall claim.

We saw that there are many philosophers and logicians who reject that necessity has anything to do with logical consequence at all, for example Philo of Megara, Diodorus Cronus, Russell, Bolzano, and Quine. Others take necessity to be a secondary feature of consequence—if Tarski is not in the first camp, he is at least in this second one. We saw that formality is similarly unsettled: Philo is also an example of a philosopher who rejected the formal nature of consequence, as are Chrysippus, many medieval logicians, Etchemendy, and Read.

Other intuitions about logical consequence are at least as disputed as these. The upshot of Chapter 1 was that the existence of a single, well-behaved intuitive conception of consequence is a myth. Beall and Restall have built enough substance into their so-called 'settled core' to make their position non-trivial but building more substance into it could reduce the admissible logics to one or even nil. Further, *monism* itself is a feature that has typically been taken to be part of the core of consequence. So Beall and Restall have stopped at an unmotivated point. This observation will be crucial in what follows.

3.5 Shapiro's eclectic pluralism

Shapiro regards eclectic pluralism as a generalization of Beall–Restall pluralism.[11] He allows all of the logics that Beall and Restall deem admissible, and many more

---

[10] Arguably, because some foundational theories might combine intuitionistic logic with strong enough principles to offset the weakness of the logic in describing classical mathematics. The question is whether those strong principles would be logical or not.

[11] Although he has a different understanding of meaning variance from them. Beall and Restall are clear that any meaning variance in their view is meta- and not object-theoretic. Shapiro holds that some pluralism can be seen as involving meaning variance at both levels.

besides. For example, he explicitly includes fuzzy logic amongst admissible ones (2014, pp. 58–9). If a logic has found an application, it is worthy of study and so falls under the scope of eclectic pluralism.

Beall and Restall are utterly clear on the criteria for being an admissible consequence relation and Shapiro is explicit that he wants to be much more permissive. Shapiro's criterion is mathematical or other interest. If a logic has found an application, or if it is of some mathematical interest, then it should be amongst the plurality of logics.

Mathematicians investigate classical structures, intuitionistic structures, classically inconsistent ones, and so on. So mathematical practice seems to require more than one consequence relation and that is enough for pluralism.

> There is no metaphysical hoop that a proposed theory must jump through. In line with Carnapian tolerance, mathematical theories have only pragmatic criteria for legitimacy. They have to be useful, or just plain interesting—as subjective as that is.   (Shapiro 2014, p. 84)

The subjectivity here is problematic. Beall and Restall can be treated as relativists about logical consequence, but their criteria remain entirely objective. Formality, necessity, and normativity are not thought to be subjective notions. Shapiro's criteria are more permissive but also, as he states, subjective.

Before moving on, let's look in a little more detail at an example of a theory cast in intuitionistic logic. This theory will illustrate Shapiro's pluralism more concretely as well as some of our more specific objections to it.

3.5.1 SIA as a test case for pluralism

In Chapter 3 of his book, Shapiro gives three examples of theories couched in intuitionistic logic which would be rendered inconsistent if the Law of Excluded Middle were added to them. The first two are Heyting Arithmetic plus an intuitionistic version of Church's Thesis and intuitionistic analysis, and the third is the lesser-known Smooth Infinitesimal Analysis (SIA). We focus on the last here, as it illustrates some interesting morals. Given that it is little known, we will take some time to spell out the details of SIA.

Like nonstandard analysis, SIA tries to put the notion of an infinitesimal quantity on a rigorous footing and develop the calculus on its basis. Just as in nonstandard analysis, the idea is that the extended real line R contains not just the real numbers but also infinitesimal quantities, the set of which may be labelled Δ. SIA differs from nonstandard analysis, however, in that all infinitesimals ϵ are nilsquare or nilpotent, that is, $\epsilon^2 = 0$, and in fact this is the definition of $\Delta = \{\epsilon : \epsilon^2 = 0\}$. Intuitively, a nilsquare is a quantity so small that squaring it turns it into 0. Another difference with nonstandard analysis is that SIA uses intuitionistic logic. For example, SIA asserts that not every nilsquare is 0, that is,

$\neg \forall \epsilon \in \Delta(\epsilon = 0)$, which is incompatible in classical logic with the assertion that $\forall x(x = 0 \vee \neg x = 0)$ (see below). As well as field axioms for R, the theory contains some more distinctive axioms, in particular the Kock-Lawvere Axiom, also known as the Principle of Microaffineness, formulated as:

> For any map $g : \Delta \to R$, there exists a *unique* $b_g \in R$ (depending on g) such that $\forall \epsilon \in \Delta, g(\epsilon) = g(0) + b_g \epsilon$.

That is, any function g is linear on nilsquares. Δ can thus be thought of as a microneighbourhood of 0 that behaves like a 'rigid rod' that cannot be bent, though it can be translated and rotated. This vivid imagery comes from J.L. Bell, who in his delightful 2008 book presents and motivates SIA and shows how elementary calculus may be simply developed within it. The reader is referred to Bell's very accessible presentation, which we'll presuppose in our critical discussion. To see briefly why SIA would be rendered inconsistent by the addition of the Law of Excluded Middle, we note that the theory proves

(i) $\forall \epsilon \in \Delta(\neg\neg\epsilon = 0)$
(ii) $\neg \forall \epsilon \in \Delta(\epsilon = 0)$

Claim (i) is proved by *reductio*. For if $\neg(\epsilon = 0)$ with $\epsilon \in \Delta$ then since R is a field, ϵ has a multiplicative inverse $\frac{1}{\epsilon}$, so that $\epsilon = \epsilon^2 \times \frac{1}{\epsilon} = 0 \times \frac{1}{\epsilon} = 0$, a contradiction. As for (ii), consider the function $g : \Delta \to R$ given by $g(\epsilon) = \epsilon$. By the Principle of Microaffineness the unique b_g for the identity function g is 1, but if $\forall \epsilon \in \Delta(\epsilon = 0)$ then b_g could also be taken as 0, which contradicts the field axiom $0 \neq 1$. With (i) and (ii) now proved, suppose $\forall x(x = 0 \vee \neg x = 0)$ so that in particular $\forall \epsilon \in \Delta(\epsilon = 0 \vee \neg \epsilon = 0)$. It follows that $\forall \epsilon \in \Delta(\neg\neg\epsilon = 0 \to \epsilon = 0)$. Using (i), we deduce that $\forall \epsilon \in \Delta(\epsilon = 0)$, which contradicts (ii). We note incidentally that (i) implies that $\neg \exists \epsilon \in \Delta \neg(\epsilon = 0)$, since $\neg\neg\epsilon = 0 \wedge \neg\epsilon = 0$ is a contradiction, a fact which will be of relevance later.

SIA is both elegant and of mathematical interest; far be it from us to deny either. We wholeheartedly concur with Shapiro that 'here, too, we have an interesting field with the look and feel of mathematics ... [that] should count as mathematics' (2014, p. 75). That said, SIA is a relatively obscure theory, which only a small handful of mathematicians have experience of. Based on admittedly anecdotal evidence—we asked several logicians we know in mathematics and philosophy departments—it seems that even among logicians, few had heard of it at the time Shapiro's book was published. In any case, if SIA is to qualify as an example of an 'interesting' and 'fruitful' theory based on intuitionistic logic abetting the pluralist case (2014, p. 67), presumably this judgement ought to be based more on the theory's logical or mathematical features than the brute subjective fact that you or I find it interesting or on the sociological fact that it has generated a slender literature. Shapiro seems to think otherwise, as he admits that his criteria are subjective.

We find the psychological and sociological criteria more or less equally uninviting. Anyone can in principle find anything interesting, for whatever reason. Subjective facts of this kind are too contingent on biological make-up, education, etc. to ground an account of logic. As for sociological facts, they too, like psychological facts, are *evidential* rather than *constitutive* of a logic's interest and utility. Facts such as which articles are published in which journals, who cites whom, and the like, are owed to a host of factors, some of which have nothing or little to do with a theory's intrinsic merits. The small number of articles, books, and talks on SIA should at most be presumptive evidence for its admissibility in the eclectic logician's eyes. We may live in an age not just of sophisters, economists, and calculators, but of citation metrics too. Still, we for one (or rather, two) are not yet prepared to make Google Scholar the arbiter of what to reasonably believe.

So which intrinsic features of a theory *should* matter for its admissibility, in the eclectic logician's eyes? It's difficult to say. Shapiro is hardly expansive on the point, insisting only that a theory ought to be interesting and fruitful. What *is* clear is that he does *not* seem to think that all that is required for acceptability is consistency. For as he notes (2014, p. 67) consistency is relative to a logic, so in order to decree that 'any logic is acceptable as long as it's consistent' we need a prior criterion of which background logic the consistency is judged relative to. The acceptability of *this* background logic is what Shapiro's eclecticism is all about; it turns on whether it's involved in interesting theories, especially ones that find application.

How does the monist see things? She can't deny the plain fact that different mathematical logics are used in different theories, even if virtually all logicians use a single (classical) logic in their metatheory. Nor should she find this logical diversity troubling: from a social-epistemological perspective, it makes perfect sense. It would be imprudent for the community of inquirers to put all its eggs in one basket. Some resources should be set aside for exploring alternatives to the mainstream logic, just as they should be set aside for exploring non-relativistic theories of space and time and non-Darwinian theories of evolution. Moreover, even if $\mathcal{L}^*$ is an implausible candidate for the correct logic, exploring $\mathcal{L}^*$-theories may be still of academic interest. Indeed, Shapiro acknowledges the observation by one of the present authors that intuitionistically framed theories such as SIA can be explored by classicists in an entirely hypothetical spirit (2014, p. 72 fn. 9). Classical logicians may wish to explore how things look from an intuitionist logician's point of view.

These responses to SIA and other theories built on intuitionistic logic in current mathematics by the monist are, of course, generic; they can be wheeled out by the monist when confronted with examples of logicians or mathematicians exploring any logic other than her preferred one. In the case of SIA, however, the monist has an even stronger hand to play. For as a means of softening monists' opposition to pluralism, SIA has two specific shortcomings.

The first point is that SIA is usually interpreted topos-theoretically. The construction has been distilled in the appendix to Bell's 2008 book,[12] so we will be very brief here. Roughly, start with the category **Man**, whose objects are smooth manifolds and morphisms smooth (infinitely differentiable) maps. Then embed **Man** into the opposite category $\mathbf{A}^{op}$ of a subcategory A of the category **Ring** of rings, and embed the latter via the Yoneda embedding into the presheaf topos $\mathbf{Set}^A$. This isn't quite a topos yet, but can be turned into one. The resulting topos models SIA. As Bell puts it, this construction guarantees SIA's consistency (2008, p. 14). In fact, to say that the topos model underwrites SIA is something of a *hysteron proteron*, since SIA was developed in the last decades of the last century by category theorists. SIA did not precede its topos-theoretic models, in the way that intuitionistic reasoning in Brouwer's hands preceded the various formal semantics for intuitionistic logic; rather, SIA fell out of topos-theoretic modelling.

As is widely appreciated, one must distinguish between a topos's internal logic and the external logic we use to reason about the topos. The internal logic is generated by considering the 'truth-values' object Ω along with the rules governing the connectives and quantifiers (given by the topos-theoretic interpretation).[13] This internal logic is intuitionistic, which is what underwrites SIA's use of intuitionistic logic. But there is no reason to suppose that the connectives and quantifiers in the topos's internal logic represent the notions we use to reason about smooth worlds. The internal logic of a topos is simply what you get when you interpret quantifiers and connectives in the specified way, in arrow-theoretic terms involving the subobject classifier Ω. Modelling logic in a topos is one thing. Using this model as a guide to the validity of implications or inferences is another.[14]

In sum, SIA is generally thought a consistent theory because it has a topos-theoretic model. Although this gives rise to an intuitionistic internal logic, the logic used to reason about topoi and categories more generally is classical. If pluralism is to be guided by practice, it has to recognize that we do not here have a self-standing invocation of intuitionistic logic. The intuitionistic theory is deemed mathematically respectable because it can be given a topos-theoretic interpretation—which is how it arose in the first place—within an entirely classical framework of reasoning about topoi.

The second point is that although SIA uses intuitionistic logic, its motivation cannot be the usual constructivist one derived from Brouwer and Heyting, and adopted by later intuitionists or intuitionist sympathizers such as Dummett. The only nilsquare constructed is 0 itself. In fact, as we saw earlier, one can prove in the theory that there are no nilsquares other than 0, i.e. that $\neg \exists \epsilon \in \Delta (\epsilon \neq 0)$. Internally, then, $\vdash \neg \exists \epsilon \in \Delta (\epsilon \neq 0)$, whereas externally as it were, the smooth infinitesimal

---

[12] See the references therein for full details of the construction.
[13] For an accessible introduction to topos logic, see Chapters 6 and 7 of Goldblatt (1984).
[14] Hellman (2006, pp. 634–5) makes a similar point.

analyst presents the theory as essentially relying on the fact that $\Delta \setminus \{0\}$ is non-empty. This point is widely recognized by commentators, from Bell (2008, p. 105) to Shapiro (2014, p. 73) and others such as Hellman (2006, pp. 629–30). As has also been noted, $\neg\forall\epsilon \in \Delta(\epsilon = 0)$ *is* provable in the theory; that does not however, change the fact that the provability of $\neg\exists\epsilon \in \Delta(\epsilon \neq 0)$ is squarely at odds with the theory's motivation. The classical inference from $\neg\forall\epsilon \in \Delta(\epsilon = 0)$ to $\exists\epsilon \in \Delta(\epsilon \neq 0)$ underwrites the smooth infinitesimal analyst's belief that $\Delta \setminus \{0\}$ is non-empty; yet the inference has to be blocked in SIA lest the system collapse into contradiction (as we saw earlier). This fundamental tension lies at the heart of SIA, the theory's formal consistency notwithstanding.

SIA is, thus, a theory at odds with the customary way of understanding it. In this sense, it fails to be a coherent interpretation of intuitionistic mathematics.[15] One could of course argue that the smooth infinitesimal analyst's belief that $\Delta \setminus \{0\}$ is non-empty is an inessential slip caused by a lingering temptation to think in classical terms. But our point still stands: although SIA is formally consistent, the usual (only?) way of understanding it is incoherent.

3.6 Metalogical reasoning

Turning away from the specific problems for these versions of pluralism, we will now put forward a general argument for any such view. Seeing that Beall and Restall's modest pluralism has a motivational problem when we consider its admissibility criteria, we may be tempted to move towards the eclectic end of the pluralist spectrum. Shapiro, for example, doesn't have this motivational problem: his admissibility criteria are sufficiently weak to allow through an awful lot more logics than Beall and Restall. He faces a different problem, however, as we will now see.

Recall from the Prologue that applied logic aims to give a theory of (cleaned-up and extended) English implication. The problem for the pluralist now is that this sort of implication occurs not only in the object language but in the metalanguage. Logical pluralists will, of course, want to assert some metalogical sentences. They may want to weigh up the virtues of different logics, they may want to discuss the properties of different logics, and they may want to offer arguments to convince other logicians to be pluralists. All of these are examples of metalogical claims. When they are engaged in such reasoning, what is their logic?

---

[15] The matter is examined in much more detail by Geoffrey Hellman (2006), who concludes that SIA should be assimilated into his modal-structuralist framework.

We will see that the answer to this question interacts with the pluralist's admissibility criteria to create a problem for eclectic pluralism. The monist is on a safer metalogical footing, since they hold that there is one true logic that captures natural-language implication, and they use this at all levels.

Beall and Restall discuss the question 'What logic are you using now?', though they do not provide a direct answer to it:

> As to *which* relation we wish our own reasoning to be evaluated by, we are happy to say: any and all (admissible) ones! Our arguments might be valid by some and invalid by others, good in some senses and bad in others. But that is not the end of the story. Once we learn that our argument is bad in some sense—for example that a verification of some premises will not itself be a verification of the conclusion—we will not necessarily *thereby* reject the usefulness of the argument. It depends, of course, on whether the given kind of verification procedure is important to the task at hand.   (Beall and Restall 2006, p. 99)

Their reply seems to be this. Assume that some piece of metalogical reasoning turns out to be, say, intuitionistically unacceptable. The significance of this point depends on whether verifiability 'is important to the task at hand'. If it is, then ruling out intuitionism is a problem. If it's not important, then there's no problem.

Shapiro's main response is the following:

> To be sure, the present study, and the discussion of the eclectic orientation generally, does involve reasoning. I hope to have convinced at least some readers of something, and, surely, there are arguments involved in that. . . . I am not aware of any inference in the present treatment that depends essentially on any of the 'disputed' inferences, such as excluded middle, disjunctive syllogism, ex falso quodlibet, or even the distributive principles. [But if there are, then] one might simply refrain from invoking excluded middle or disjunctive syllogism anywhere in the informal treatment, sticking to argument forms that are what I call 'super-valid'.   (Shapiro 2014, pp. 168–9)

Shapiro's main response is that none of his arguments rely on controversial inferences. The uncontentious rules that everyone would accept he calls 'super-valid'. (As we will come to in §3.8.4, Shapiro also has a second response up his sleeve.)

It's somewhat surprising that these authors are not clearer and more definite about which logic their respective books use. Still, they suggest some options, and we can work on their behalf by exploring them and others. The options suggested in these quotations are two, to which we will add another one by Shapiro and a related one that might appear promising. The first is Shapiro's response that pluralists should use only rules common to all the true logics (§§3.8.1–3.8.2).

The second is Beall and Restall's response that we should use the logic correct in this particular domain—that of metalogical reasoning (§3.8.3). The third is simply to pick some logic or other, which as we will see is the second of Shapiro's responses (§3.8.4). The fourth and final option is to offer an argument in each true logic (§3.8.5). An altogether different sort of response is to suppose that no logic is being used when arguing for pluralism, but only non-deductive reasoning; this suggestion will be examined in §3.9.1.

We note in passing an *ad hominem* point that we do not want to make much of, but which is indicative of a broader moral. Several of the arguments Beall and Restall advance throughout their book rely on controversial inferences, unsurprisingly since we all rely on such inferences frequently. The authors for example caution against arguments that would fail to convince the intuitionist, but rely on such arguments themselves in several places. We give two examples from a range of possible ones. First, they rely on an instance of excluded middle when justifying that classical consequence holds of necessity (2006, p. 40). In particular, they rely on the premise that there are either enough objects (including *abstracta*) to construct classical models, or there are not. Second, they do so again when arguing that the necessary truth-preservation account of consequence might fail to be formal in a certain way (2006, p. 42). In this instance, they rely on the premise that either there are necessary existents or there are not. As with Beall and Restall, and as might be expected, there are several places where Shapiro uses controversial inferences. Sticking with the same example of excluded middle, we find Shapiro using it, for instance, on pp. 168–9 of his book. He first supposes that he doesn't need to pick one metalogic, then that he does, and argues that in either case there's no problem. The same law is invoked on p. 109 when he discusses different views on stages in Kripke structures. Our concern is with pluralism generally, so we do not want *ad hominem* arguments of this sort to take centre stage. But they serve as a warming-up exercise for our more general point. Setting stringent constraints on the metalogic can be fatal for the pluralist's argument.

3.7 Logical nihilism

Shapiro's talk of 'super-valid' inferences naturally raises the question: which, if any, inferences are super-valid? It is well-known that many laws of classical logic have been contested, for example the law of excluded middle and the law of non-contradiction. So are there *any* uncontroversially correct logical laws?

Logical *nihilism* is the view that there are not. Even if logical nihilism is false (as we believe), the lesson from it is that few rules are uncontroversial and thus very few arguments super-valid. This is a problem for a permissive pluralist such as Shapiro, since his metalogic includes only the super-valid inferences, of which there seem to be none.

There is a small recent literature on logical nihilism and it will be instructive to review it briefly, since it may be unfamiliar. Aaron Cotnoir (2018) expresses two versions of the view, the one most relevant for present purposes being:[16]

Logical Nihilism There are no constraints on natural-language inference; there are always counterexamples to any purportedly valid forms.

Although we are not logical nihilists, we wish to extract an important insight from it: disagreement about logical principles is virtually unlimited. Let's consider some instances of controversy.

In Gillian Russell's (2018) discussion of logical nihilism, she considers the rules of conjunction-elimination ($\wedge$E) and identity (ID), as these are often thought 'to be the safest and most secure laws of logic' (p. 128):

$$\frac{A \wedge B}{A} \ (\wedge E) \qquad \frac{A \wedge B}{B} \ (\wedge E)$$

$$\frac{A}{A} \ (ID)$$

Russell's strategy is to find counterexamples to ($\wedge$E) and (ID) using self-referential premises or conclusions. For example, consider a sentence whose truth-value depends on whether it is in the scope of a conjunction. Russell (2018, p. 128) considers 'con-white', a predicate whose extension is that of 'white' when in the scope of a conjunction, and empty otherwise. Then she puts forward the argument:

$$\frac{\text{Snow is con-white and grass is green}}{\text{Snow is con-white}}$$

The premise is true because it has two true conjuncts: grass *is* green and snow *is* con-white here because it is in a conjunction. The conclusion is false because snow is *not* con-white when the sentence is not part of a conjunction. So the argument is invalid, and yet is an instance of ($\wedge$E).

The technique can be extended to (ID) in the obvious way. We need a new predicate 'prem-white', whose extension is the same as 'white' when in the premise of an argument, and empty otherwise. The following instance of (ID) is invalid:

$$\frac{\text{Snow is prem-white}}{\text{Snow is prem-white}}$$

---

[16] Cotnoir's (2018) other version of logical nihilism is: 'There's *no* logical consequence relation that correctly represents natural-language inference; formal logics are inadequate to capture informal inference.'

The premise is true, since snow *is* prem-white when the sentence is a premise, but the conclusion is false. Like (∧E), (ID) fails to be exceptionless. Clearly, Russell's method will generalize and, if legitimate, can be used to show any purportedly exceptionless natural-language argument form to have invalid instances.[17] Aristotle is often interpreted as having denied the validity of (ID) on the basis that arguments involving it are not proper arguments: 'If, however, the relation of B [the premise] to C [the conclusion] is such that they are identical, or that they are clearly convertible, or that one applies to the other, then he [who reasons thus] is begging the point at issue' (*Prior Analytics*, Bk II xvi, 64b28–65a26).

Russell has provided evidence for Logical Nihilism; roughly, that there are counterexamples to every English argument form. And even if responses are available to her particular cases, this is a thought that many have held. She cites Williamson in support: 'All major logical principles have been rejected on metaphysical grounds' (2013, p. 146). Indeed, Williamson (2011, pp. 499–502) discusses (though does not endorse) another alleged counterexample to (∧E). It involves someone who accepts three-valued weak Kleene logic and takes 'true' and 'undefined' as designated values. Then the argument $P \wedge Q \therefore P$ has an undefined premise and a false conclusion when P is false and Q undefined.

James Cargile discusses (∧E) further. He considers the example 'I'm happy and I'm sad; *so* I'm happy' and writes:

> what the practicality of English is incompatible with is mathematically precise pairing of syntactic structure with logical functions   (Cargile 2010, p. 92)

He believes that the practicality of English entails that even rules such as (∧E) have counterexamples and he thinks we wouldn't normally take the ambivalent 'I'm happy and I'm sad' to entail that 'I'm happy'. He adds that

> what the practicality of English is incompatible with is the view that validity is entirely due to a logical form inherent in the sentences of the argument.
> (Cargile 2010, p. 113)

and that the rules 'were "made to be broken" by poets, or philosophers or simple people needing charitable hearing' (2010, p. 92). Strawson (1952) expresses similar

---

[17] Russell's deviant 'con-' and 'prem-' predicates involve self-reference. The first argument is essentially 'This sentence is true when part of a conjunction and grass is green; *so* this sentence is true when part of a conjunction'. This version brings out the self-reference implicit in Russell's argument. Some may be suspicious of the (essential) self-reference involved and ban such instances of (∧E) on this basis. This was of course the other Russell's (1908) response to the Liar Paradox, and his ramified type theory rendered Liar-like sentences ill-formed. Along with most, we doubt that self-reference is always in and of itself a problem. There are many perfectly benign forms of self-reference in natural language (e.g. 'This sentence is short') or logic (e.g. diagonal sentences).

thoughts. In his book, he argues that even non-contradiction has counterexamples in English:

> 'It is and it is not' may be held to exemplify the verbal pattern 'P and not-P' and yet may be used to give a perfectly consistent answer to a question (e.g., 'Is it raining?')   (Strawson 1952, p. 42)

Finally, Vann McGee (1985) offered an alleged counterexample to the seemingly sacrosanct *modus ponens*:

1. If a Republican wins the election, then if it's not Reagan who wins it will be Anderson.
2. A Republican will win the election.
∴ 3. If it's not Reagan who wins, it will be Anderson.

The thought is that, ahead of the 1980 presidential election, Republican Reagan was decisively ahead of Democrat Carter, with other Republican Anderson a distant third. In light of this polling, we may well believe 1. and 2. but not 3. The lesson is supposed to be that *modus ponens* is not a wholly reliable rule of inference.[18]

In sum, Shapiro has invited us to consider what the 'super-valid' inferences are. We have seen good evidence that there are very few such inferences. At least Gillian Russell, Strawson, Cargile, and McGee have questioned these inferences that are often taken to be exceptionless.[19] Admittedly, few authors have questioned (ID) but if that is the only rule the pluralist has, they are, in a very clear sense, only going to convince those who are already pluralists. And logics that deny any given one of these rules could, and in many cases have been, constructed, which is proof of their mathematical interest. Now we are monists who believe that there *are* exceptionless formal inferences, and our aim isn't to defend logical nihilism. The point rather is a dialectical one: if the test for whether an inference is 'super-valid' involves looking for near-consensus amongst logicians, then the test is likely to deliver no rules whatsoever. All logical rules have been disputed.

3.8 A problem for logical pluralism

Applied logic provides a theory of implication. The problem for the pluralist now is that this sort of implication occurs not only in the object language but in the metalanguage. Whenever they offer an *argument* for some metalogical claim, they

---

[18] One line of response would be to deny that McGee's conditionals are genuinely indicative. Indeed, we might challenge the indicative/subjunctive distinction (see Dudman 1988).

[19] Estrada-González (2011) also discusses similar cases.

need a logic. Beall and Restall, and Shapiro, seem to hold that their metalogical arguments don't rely on controversial deductive inferences but they do rely on *deductive inferences* and those are logical.[20] We will now consider the various possibilities for the pluralist.

3.8.1 All true logics

Our argument against the pluralist crucially involves metalogical reasoning. Let's start with Shapiro's eclectic perspective which, as we have seen, is enormously permissive: any logic of intrinsic technical interest or that has had a fruitful application is allowed. If, as Shapiro claims, the metalogic should use only those rules accepted by all of the logics canvassed, there will be very few, if any, available. Letting a thousand flowers bloom leads to the threat of logical nihilism.

Perhaps the eclectic pluralist could reply that the weakest logics considered are not fruitful or interesting enough to warrant consideration. This is a difficult line to maintain. Shapiro uses the term 'super-valid' to describe the uncontroversial inferences, but the discussion of logical nihilism shows that even seemingly sacrosanct inferences are to some extent controversial to Russell, Strawson, Cargile, McGee, and others.

A pluralist would need to further explain the notion of 'super-valid' to pursue this response. Or they may reply that they never intended to appeal to *all* logicians. Granted, this is Shapiro's aim (at least, all logicians whose favoured logics are of formal interest or have a fruitful application), but many pluralists are content with some small subset of all logics. Beall and Restall, of course, are an example of such modest pluralists.

But, as we argued in Chapter 1 and briefly summarized in §3.3.1, Beall and Restall's modesty is unmotivated. Very briefly, their test for whether a feature should be considered settled is historical (e.g. 2006, p. 14 and p. 29). If we look to the tradition of writing about consequence, we apparently find some features— necessity, formality, and normativity—settled, and others—axiomatizability, *a priori* knowability—unsettled. When we consider the tradition, however, no such picture emerges. Rather, we find a thoroughly unsettled picture that couldn't possibly motivate a modest variety of pluralism. It would certainly be hard to argue that some feature is more part of the tradition than the idea of monism itself: so if anything is a settled feature, it's the thought that there is one correct logic.

To sum up: in light of its weak admissibility criteria, eclectic pluralism leads to an unsatisfactory metatheory. Beall and Restall's modest pluralism fails due to their mistaken claims about the settled core. It is open to the modest pluralist to find

---

[20] We consider the possibility of their using non-deductive inferences in §3.9.

some other subset of logicians to convince to be pluralists. But whatever subset is chosen must be motivated, and we have found no other motivation on the market.

3.8.2 The single correct metalogic

We now turn to the idea that there is a single logic appropriate for metalogical reasoning. If so, that is evidently the logic a pluralist should use to convince us of pluralism.

To spell this out somewhat, imagine that our pluralist sees some domain D, distinct from that of metalogical reasoning, as admitting at least two equally acceptable distinct logics, $\mathcal{L}_1$ and $\mathcal{L}_2$ say. (For the sake of concreteness, you can if you like imagine D to be the domain of empirical reasoning, or some narrower subset thereof, e.g. quantum-mechanical reasoning.) When it comes to metalogical reasoning, though, the pluralist upholds a single correct logic, $\mathcal{L}^M$. This is a form of pluralism, since there are some arguments in domain D about whose validity $\mathcal{L}_1$ and $\mathcal{L}_2$ disagree, yet neither logic is to be preferred to be the other. But it seems that such a pluralism is not susceptible to our metalogical argument, as it prescribes a *single* logic—$\mathcal{L}^M$—for metalogical reasoning.

The first thing to say about this combination of views is that it is not, so far as we know, instantiated. True, the reason we cover it here is that it was prompted by Beall and Restall's claim that the right logic is the one suitable 'to the task at hand'. But they say little beyond this hint, so it is hard to ascribe it to them. More generally, the pluralists out there don't think there's anything special about the domain of metalogical reasoning that shields it from their pluralist arguments. It seems to be entailed by all they say that pluralism extends to this domain too. This deals with the pluralism defended in the literature and throws down the gauntlet to any future ones.

The second point is that this form of monism-in-the-metalanguage-but-pluralism-in-some-part-of-the-object-language courts inconsistency. Take a D-argument about whose validity logics $\mathcal{L}_1$ and $\mathcal{L}_2$ disagree, say the argument with premise A and conclusion B.[21] Without loss of generality, suppose that A entails B according to $\mathcal{L}_1$ but not according to $\mathcal{L}_2$. So when thinking about D and reasoning on the assumption that A, an $\mathcal{L}_1$-using logician is legitimately entitled to infer B; but by the same token, an $\mathcal{L}_2$-using logician is legitimately entitled not to infer B from A when considering whether to do so. Or so says the species of pluralism under scrutiny. Now let's move to the metalanguage, and consider the claims 'A is true' and 'B is true'. Does 'A is true' entail 'B is true'? The answer is either yes if $\mathcal{L}^M$ condones it, or not yes if $\mathcal{L}^M$ does not—in each case $\mathcal{L}^M$ is the arbiter.[22] But if yes, the answer is in tension with $\mathcal{L}_2$'s verdict that one cannot

---

[21] The assumption of a single premise is for simplicity only.
[22] We are careful here not to equate 'no' with 'not yes'.

logically infer B from A; and if not yes, it's in tension with $\mathcal{L}_1$'s verdict that one can logically infer B from A. These tensions can be converted into an outright contradiction by specifying disquotation principles from which it follows that the argument

'A is true' therefore 'B is true'

is valid if and only if 'A therefore B' is valid.

Now we recognize that the argument in the last paragraph is not iron-clad. One could argue that there are no disquotation principles of the sort relied on to be had. As logicians know full well, naive truth theory cannot be consistently combined with classical logic,[23] yet our argument seems blithely to assume both. But again, here we risk tilting at windmills. For although many logicians have considered restricting disquotation principles for sentences containing semantic vocabulary, they usually strive to respect at least non-semantic instances of disquotation. And virtually by definition, neither of the sentences involved in the inference 'A therefore B' is semantic, since A and B are drawn from domain D which *ex hypothesi* is the domain not of metalogical reasoning but of some form of non-semantic inquiry. For this reason, we take plausible versions of monism-in-the-metalanguage-but-pluralism-in-some-part-of-the-object-language to be vulnerable to the argument just sketched.

3.8.3 Some arbitrary true logic

In the same passage where Shapiro proposed what we called his main response, he also suggested an alternative one. He wrote: 'Suppose, however, that in order to articulate and defend the present eclectic orientation, I have to adopt and use a particular logic. One option would be to adopt classical logic without apology for the philosophical enterprise at hand' (Shapiro 2014, p. 169). This alternative response, the third in our list, is that a pluralist could simply pick a particular logic in which to present the argument. This will be a true logic, since if it is correctly used in some domain or other—here, metalogical reasoning—then by definition it is true. Shapiro mentions classical logic, but does not suggest there might be anything special about it in this regard, so it could just as well be any true logic.

To begin with, we note that there is no sign that either Beall and Restall or Shapiro proceed in this way. Shapiro himself undercuts the motivations for the 'pick an arbitrary logic' approach as soon as he introduces it: 'An advocate of this approach might even provide an argument that this is the appropriate logic for this

---

[23] See Halbach (2014) or Beall *et al.* (2018).

particular enterprise, although I cannot imagine what such an argument would look like and I would think that its premises might be just as controversial as the underlying eclectic thesis itself' (2014, p. 169).

There is also overwhelming pressure for this response to turn into the one examined in §3.8.4, that a pluralist should instead offer an argument for their position in *each* true logic. For first, as Shapiro realizes, it seems arbitrary to choose some particular logic for this job—what is the rationale for couching the pluralist argument in true logic $\mathcal{L}_1$ and not true logic $\mathcal{L}_2$? The pluralist cannot reply that $\mathcal{L}_1$ but not $\mathcal{L}_2$ is the correct logic for metalogical reasoning—that would simply take us back to §3.8.2.

Second, this approach cannot hope to dissolve the disputes between monists of various stripes. The promise of doing so is part of the appeal of contemporary pluralism and arguably its *raison d'être*. If there is only a classical-logic-using argument for pluralism, for instance, then intuitionists would lack an intuitionistically acceptable argument that intuitionistic logic is a true logic.

Third, a straightforward amendment of the argument in §3.8.2 shows that the true metalogic must be in harmony with the true object-level logics. If 'A therefore B' is valid according to one true logic $\mathcal{L}_1$ but not according to another true logic $\mathcal{L}_2$, then this divergence must be reflected in the metalanguage. Via semantic ascent (T-principles), the claim that 'A is true' entails 'B is true' must be true according to $\mathcal{L}_1$ but not $\mathcal{L}_2$. So all true logics get lifted to the metalevel. In sum, the answer 'pick one logic' collapses into another one, 'each logic', which we turn to next.

3.8.4 Each true logic

The fourth and final option is not suggested by any pluralist we know but, as we saw, leads on from the third: offer an $\mathcal{L}$-argument for each true logic $\mathcal{L}$.

Against this, we first begin by once more noting that this is not how either modest or eclectic pluralists proceed. They simply offer a single argument rather than a series of arguments, each from the perspective of the different logics they espouse. In fact, it would be quite odd to suppose that there isn't a single underlying argument for pluralism, but that it must be recast in different ways from different perspectives. By far the most natural thing to say is that if there is a good argument for pluralism, then that same argument should be frameable in any true logic—and so much the worse for any logic that does not allow for its expression.

The most telling point against this approach, however, is similar to an earlier one. For the eclectic pluralist, the weakest logics in question are extremely weak. So, even though the argument for pluralism may be different in each logic, there must be such an argument in each of them. For the pluralist, it would have to be the case that their argument didn't essentially rely on any of the inferential rules employed. But, as we saw in our discussion of logical nihilism (§3.7), very few logical principles have gone unquestioned. Even ($\wedge$E) and (ID) have come under recent scrutiny. The pluralist may at this point want to dismiss such hopelessly weak logics and focus on stronger cases. But now the other part of our argument comes into play: is there another motivated stopping point other than monism? In other words, the pluralist could—like Beall and Restall—think that this just shows we should not consider every logic. Instead, we should limit our attention to some better known logics, such as classical, relevant, and intuitionist. But, as we argued earlier, such a stopping point is unmotivated.[24]

3.8.5 Summary

The pluralist will want to do at least some metalogical reasoning. What inferences can they make here? We considered four answers: the inferences common to all the true logics (§§3.8.1–3.8.2); those in the single correct logic for metalogical reasoning (§3.8.3); those in some arbitrary true logic (§3.8.4); finally, those in each true logic (§3.8.5). We saw that none of these is tenable.

It will be instructive to briefly summarize our response to the first answer, as its moral is very general. How many logics pluralists canvass depends on their admissibility criteria. If, like Shapiro, they endorse an eclectic pluralism with extremely weak admissibility criteria, nihilism in the metalogic threatens. If, like Beall and Restall, they endorse a modest pluralism with stronger admissibility criteria, they may have a workable metatheory. But the problem now is that the criteria are unmotivated: the so-called 'tradition' delivers nothing like what they need. We have seen no pluralism with admissibility criteria that are motivated *and* deliver a reasonable metatheory. This is not, of course, to say such criteria cannot be delivered, but we know of no such examples and it is far from clear that such criteria are possible.

3.9 Two further issues

We have seen that there is no satisfactory pluralist metalogic—no satisfactory logic in which the pluralist might propound their argument(s) for pluralism. We round off our discussion by considering two further issues. The first is the idea that the

---

[24] Shapiro (2014, §7) suggests one other pluralist response, which we might call 'pluralism all the way up'. The thought seems to be that the pluralist can endorse all of their logics in the metatheory as well as in the object theory. Such a response has some appeal: after all, if some logic cannot be used for metatheoretic reasoning, then there is a domain to which it does not apply. This section's 'each logic, not every logic' response applies here. The metatheoretic reasoning in question should be possible in each of the logics canvassed, perhaps using a different argument each time. But when the pluralist is *eclectic*, these logics will include extremely weak logics in which any sophisticated reasoning is impossible.

pluralist's argument need not be deductive (§3.9.1). The second is whether the problem here raised for pluralism affects monism in any way (§3.9.2).

3.9.1 Non-deductive reasoning

One response to our argument is to deny that metalogical reasoning needs to be *deductive*. Perhaps *non-deductive* arguments would be more appropriate. Indeed, Shapiro's argument for eclectic pluralism is largely abductive and he notes that '[t]he abductive aspects of the thesis, such as the viability of non-classical theories, are more plausible targets for criticism.' (2014, p. 169).

While some metalogical arguments, such as those for logical pluralism, might be non-deductive, it is unlikely that all such arguments will be. The metalogical reasoning might involve discussion of which logics have certain properties, or which is the best for a certain purpose. It is implausible that the arguments for all such claims can avoid deductive reasoning altogether. At one point Shapiro comments, somewhat wishfully, that his arguments are 'perhaps not deductive at all' (2014, p. 169). But that is clearly not the case, quite apart from the relatively sophisticated example mentioned in §3.6 (where we also presented examples from Beall and Restall's writings of inferences contested by some of the logics they champion). At an abstract level, the most obvious such example is that anyone who infers pluralism (an existential claim) from examples of true logics (instances) deploys a deductive inference—existential generalization. More generally, if we look outside logic and mathematics to, say, natural science, we find an abundance of abductive and inductive arguments, but equally many straightforwardly deductive ones. Even if the overall shape of a scientific argument is abductive, many of its parts contain deductive inferences, many so obvious or so trivial that they usually escape our notice. It is just as easy to underestimate how much deductive inference takes place outside mathematics and logic as it is to overestimate it.

We can go even further than this. All analyses of abductive reasoning we know of rely on deductive reasoning. The best explanation of some evidence or data *entails* it (possibly in combination with some auxiliary assumptions). It should also, whenever possible, *entail* new claims that can be subject to confirmation. The best explanation must also *cohere* with other hypotheses we accept, a necessary condition for coherence being *consistency*. And so on. Now, we are not advocating a simple hypothetico-deductive model of abductive reasoning, which, though it has much to recommend it, faces equally formidable problems.[25] Our point is rather that *any* account of abductive reasoning that could apply in the case at hand will make use of logical consequence, be it some version of the

---

[25] See for example Hempel and Oppenheim (1948) and Godfrey-Smith (2003, ch. 13).

hypothetico-deductive model, or the unification account (Friedman 1974; Kitcher 1981, 1989), van Fraassen's pragmatic account (1980, ch. 5), or Lange's account of non-causal explanation (2017, part I), or any other.[26] All such models of abductive reasoning start with logical reasoning as given and build upon it. When we look under the bonnet of abductive reasoning, we find that it relies heavily on logical reasoning.[27]

3.9.2 A problem for monism?

Finally, the pluralist could try to turn the tables on us. What metalogic can the *monist* use?

The answer is straightforward: the monist can, will, and should just use the logic they believe to be the single true one—what else? A classical-logic monist, for example, can justify their belief that classical logic is the one true logic by giving an argument that uses classical logic. In other words, the monist faces no problem of metalogical reasoning.

It is important to distinguish this (easily met) challenge from two different worries. The first worry is whether the monist can offer a *dialectically* persuasive argument for their view. In other words, if the monist wants to convince, say, the classicist, the intuitionist, and the relevantist, must they not also confine themselves to deductive inferences accepted by all three? We agree that this is an important challenge. But it's clearly different from the one we have been focusing on in this chapter. Our question has been how to justify a logic in its own terms—this is what we claimed was a problem for pluralism. Whether you can justify your preferred logic in terms acceptable to all-comers—including an assortment of logicians you would regard as deviant—is another question. We suggest that the answer to this latter question is no, since, as we have seen, some hypothetical logicians may espouse only 'super-valid' inferences in Shapiro's sense, which are vanishingly few. It is very hard to persuade such logicians of anything they do not antecedently believe.

A second worry has to do with the admissibility of the one true logic. A true logic must, amongst other features, be admissible. We saw that, because admissibility is so easily had from the eclectic perspective, logical nihilism threatens. The modest pluralist with stronger admissibility criteria, such as Beall and Restall's, is not threatened by nihilism; but the problem for *them* is that their admissibility criteria are unmotivated. The challenge for the monist is to provide admissibility criteria, in such a way as to avoid the Scylla of nihilism and the Charybdis of modest (or even eclectic) pluralism. Parts II and III of this book will be principally concerned

---

[26] The same could be said of more empirical models, such as Salmon's causal-mechanistic account (1998) or Woodward's manipulationist one (2003).

[27] See Woods (2019) for more on how abductive reasoning relies on *deductive* reasoning.

with arguing that the one true logic is maximally infinitary, but they will also set out in passing some admissibility criteria for it (see especially Chapters 6 and 7).

3.10 Conclusion

In summary, logical pluralism is an unstable position. The logical pluralist must carry out some metalogical reasoning. This reasoning will take place in a logic, but which one? We have seen that there is no good answer to this question, either from Shapiro's eclectic perspective or from Beall and Reall's modest one. The eclectic perspective leaves us with very few rules, if any at all. The modest perspective must rest on a motivated reason for endorsing some logics and no others. Beall and Restall have not motivated their stopping point, and although there could be another, none has been suggested so far, so that we are doubtful this line can be convincingly drawn.

Logical pluralism, then, is an unattractive view. It is also an unintuitive one: when we look to how logic is used in science, mathematics, the law, and virtually any other domain, it is usually thought that there is a fact of the matter as to whether a given argument is valid. Logical monism respects this by putting forward a single correct logic. Logical pluralism disrespects this and is unstable. The crucial task for the monist is, of course, to specify which of the many logics on the market is the one true logic. This is the question to which we now turn.

Appendix: metalogical meaning variance

We have seen that Beall and Restall want to avoid meaning variance at the object level: they want the logical constants to have the same meaning in all of the logics they canvass. On the other hand, they allow meaning variance in the metatheory, for instance, they allow that 'logical consequence' can change meaning between logics. In this appendix, we will explore some of the issues involved here, and argue that Beall and Restall face significant challenges.

Recall Beall and Restall's insistence that their logical pluralism is not Carnap's: 'For us, pluralism can arise *within* a language as well as between languages' (2006, p. 79). For Carnap, there are different, equally correct, logical consequence relations in different languages: '*In logic, there are no morals.* Everyone is at liberty to build his own logic, i.e. his own form of language, as he wishes' (1934, §17). If their pluralism is to be substantially different from Carnap's, they must not be changing the language as they move between logics. In particular, they must not be changing the meanings of the logical constants.

Graham Priest, in an unpublished 1999 manuscript quoted by Beall and Restall (2006, p. 97) raises a concern about meaning variance: 'we may give either intuitionist truth conditions or classical truth conditions.... But the results are not equally legitimate. The two give us, in effect, different theories of vernacular connectives: they cannot both be right'. The thought is that we can give the meanings of the logical constants by their truth conditions, but that the truth conditions of classical and intuitionistic negation are different and so they have different meanings.

At first glance, the truth conditions that Beall and Restall give (see e.g. 2001, p. 8; 2006, p. 97) for the three negations do seem different. Although Priest only mentions classical and intuitionistic logic, we also include Beall and Restall's negation clause for relevant logic:

Classical $\neg A$ is true in a Tarskian model M if and only if A is not true in M.

Intuitionist $\neg A$ is true in a construction c if and only if there is no construction d extending c in which A is true.

Relevant $\neg A$ is true in a situation s if and only if for each situation t compatible with s, A is not true in t.

Beall and Restall believe that there is no tension between these three negation clauses. We will go through the three instantiations of 'cases$_x$' in turn.

First, *models* are complete in the sense that every sentence of the language is either true or false, and consistent in the sense that no sentence is both true and false.

Second, *constructions*, or 'stages' as Beall and Restall sometimes call them (e.g. 2006, p. 62), are consistent but potentially incomplete, in the sense that some sentence of the language may have no truth-value: it may be neither true nor false. A construction d extends another construction c if, and only if, every sentence that receives a truth-value in both receives the same truth-value in both and there is some sentence that has no truth-value in c but receives a truth-value in d. If a construction c decides every sentence of a language, that is, it is a 'final construction' in Beall and Restall's terminology, then c is a model.

Third, *situations* are potentially incomplete and inconsistent. For Beall and Restall, one situation is compatible with another if, and only if, every sentence that receives a truth-value in both receives the *same* truth-value in both. Models are a *proper* subset of the situations, namely the complete and consistent situations.

As Restall notes in a solo paper on the subject, the accounts 'agree where they overlap' (2002, p. 438). The models are all and only the complete and consistent situations, and all and only the complete constructions. And Beall and Restall think this lack of conflict is enough for sameness of truth conditions, and so meaning:

> [Priest's objection] would be a fatal objection to pluralism if it were the case. How can each clause for negation be *equally* accurate? The clauses can both be equally accurate in exactly the same way as claims about a thing can be equally true: they can be equally true of one and the same object simply in virtue of being *incomplete* claims about the object. What is required is that such *incomplete* claims do not *conflict*, but the clauses governing negation do not conflict.... Each clause picks out a different feature of negation.   (Beall and Restall 2006, p. 98)[28]

---

[28] In an earlier paper, a different argument against meaning variance can be found:

> How can any number of claims about a thing be equally true claims about that thing? They can be equally true simply in virtue of being *incomplete* claims about the object in view. If JC says that Graham Priest is a philosopher, if Greg says that he is a Marxist, and if someone else says that he is accomplished at karate, then it does not follow that we are not all correct. None of us, however, has told the *complete* story of Graham Priest. Each of us has described one of his features. The descriptions can be (and are) jointly true.   (Beall and Restall 2001, p. 8)

This argument is entirely unconvincing: various different claims about an object can of course be true on account of incompleteness, but Priest's argument was about *meaning*. That the various claims about Priest are true in no way entails that 'Graham Priest' *means* Being a philosopher, Being accomplished at

The idea is that negation is the same phenomenon in each logic, but that different aspects of negation are used in each instance. '¬' means the same in each logic, but it ranges over different kinds of cases: models in the classical case, constructions in the intuitionist, and situations in the relevant. So far, this is no more than a suggestion as to why Beall–Restall pluralism avoids meaning pluralism. To be convincing, they need to provide one clause for the truth-conditions of negation that can plausibly be taken as giving the meaning of '¬' and with which classical, intuitionist, and relevant logicians can agree.

Hjortland (2013) distinguishes two meaning variance theses discussed by Beall and Restall:

(A) The meaning of 'valid' varies across logical theories;
(B) The meaning of some logical connective ⊗ varies across logical theories.

Beall and Restall endorse (A)-style meaning variance but deny (B)-style. Whether this combination of views is possible depends on how 'meaning' is understood. We'll consider three interpretations—two found in Beall and Restall and one of our own—and argue that none of them delivers what they want. We first present the interpretations, and then offer our criticism.

First, let's take 'meaning' to be *sense*. Beall and Restall consider this: 'We hold that there is more than one sense in which arguments may be *deductively valid*, that these senses are equally good, and equally deserving of the name *deductive validity*' (2001, p. 1).

Beall and Restall now require that the *sense* of, say, negation does not change as they move between logics. They offer the following biconditional to perform the required task:

[NEG] ¬A is true in x if and only if A is not true in x.[29]

If Beall and Restall offer NEG as giving the meaning of negation, they in effect avoid the differences between the classical, intuitionist, and relevant negation clauses by abstracting away from the specific cases featuring in each and replacing them with a variable to yield NEG, which provides the single meaning of negation, or so they claim.

One problem with this approach is that straight substitution of 'situations' or 'constructions' for 'x' in NEG does not yield the desired truth conditions. So Hjortland (2013, p. 9) suggests, though does not endorse, the following schematic biconditional:

[NEG*] ¬A is true in a case x if and only if, for every case x' such that x is R-related to x', A is not true in x'.

Here, by varying the R-relation, the different negations under discussion can be semantically characterized on different precisifications. This is the most plausible schematic biconditional that we have found to serve Beall and Restall's purposes.

Second, Restall elsewhere understands the (A)-style meaning variance thesis in terms of both *sense* and *reference*: 'there are at least two distinct relations of logical consequence and not simply two different relations in intension, but two distinct relations in extension' (Restall 2002, p. 426). The thought here is that the different precisifications of GTT yield notions of logical consequence that differ in both sense and reference.

A third understanding not discussed by Beall and Restall ties meaning to what Kaplan (1989) calls *character*; this possibility is discussed by Russell (2008, p. 604). The character of an expression is what determines its content in different contexts, which Kaplan models

---

karate, or Being a Marxist. Their argument lends no support to the view that the three negations have the same *meaning*.

[29] Beall and Restall (2000, p. 481; 2006, p. 51).

as a function from a context to a content. In the case of GTT, the context is a class of cases and the content is a consequence relation. On this view, despite, for example, classical and intuitionistic consequence differing in sense and reference, they share a single character given by GTT: a function from cases to extensions.

There are, therefore, three ways in which (A)-style meaning variance can be understood and, correspondingly, three ways in which the (B)-style can be understood: in terms of sense, reference, or character. Beall and Restall require an understanding of meaning on which (A)-style meaning variance holds—'valid' differs in meaning across logics—but on which (B)-style fails—the meaning of the connectives does not differ across logics.

Clearly, if we understand meaning in terms of sense, Beall and Restall have both (A)- and (B)-style meaning variance. By their own lights, 'validity' differs in sense between logics because of the different sorts of case involved, so the meanings of the connectives must likewise differ in sense, since they too are defined in terms of different sorts of cases. So, with 'meaning' understood in terms of sense, (A)-entails (B)-style meaning variance.

Similarly, if we understand meaning in terms of reference, Beall and Restall have both (A)- and (B)-style meaning variance. The kinds of case involved in different definitions of 'validity' for the different logics determine different extensions of 'validity' and so do those for the logical connectives. So Beall and Restall cannot coherently endorse (A)-style meaning variance while simultaneously denying (B)-style.

The best option for Beall and Restall, then, may be to understand 'meaning' in terms of *character*: GTT captures the character of validity and NEG* captures the character of negation.[30] If Beall and Restall understand meaning in terms of character, however, then (A)- and (B)-style meaning variance still stand or fall together. If we allow that NEG* captures the character of negation, and that meaning is exhausted by character, then the meaning of negation remains constant between a plurality of logics, so (B)-style meaning variance is avoided. But, (A)-style meaning variance is also avoided, since the character of validity similarly remains constant across the same plurality of logics. So we have still not been given an understanding of meaning on which (A)-style meaning variance holds but (B)-style does not. In short, however we understand meaning, (A)-style meaning variance, which Beall and Restall endorse, entails (B)-style meaning variance, which they reject.

Beall and Restall's claim, therefore, that they can hold (A)- but not (B)-style meaning variance is indefensible: they have to give up either their acceptance of (A)-style or rejection of (B)-style meaning variance. Given their insistence that their pluralism is not that of Carnap, we believe the rejection of (B)-style meaning variance to be the more important claim of Beall–Restall pluralism. Without it, their pluralism collapses to Carnap's. They should therefore abandon their (A)-style meaning variance and accept that 'validity' does indeed mean the same in each of the logics they endorse. The most promising way to do this is to identify meaning with character.

In summary, Beall–Restall pluralism avoids (B)-style meaning variance only if (i) the meaning of logical connectives is identified neither with sense nor reference but with character in the sense of Kaplan; (ii) an appropriate schematic biconditional can be given for each connective; and (iii) this biconditional has some independent justification away from serving the purposes of Beall–Restall pluralism.

We leave (i) and (ii) as claims that Beall and Restall must defend for their position to be plausible. But we have argued against (iii) in the case of negation. Just as their settled core of consequence is unmotivated, so is their schematic biconditional giving the meaning of negation.

---

[30] See Erik Stei (2020) for a different defence of pluralism using Kaplan's character.

PART II
THE L∞G∞S HYPOTHESIS

4

The L∞G∞S Hypothesis

ἐν ἀρχῇ ἦν ὁ λόγος
In the beginning was the Word
(John 1:1)

4.1 Introduction

In Part I, we argued that there is one true logic. In Part II, we now shift gears and consider which logic this is. Logical tradition of the past hundred years gives first-order logic a privileged role.[1] So let's start there.

Over the first half of the twentieth century, first-order logic emerged as logicians' and mathematicians' foundational logic of choice.[2] In philosophy, first-order logic's greatest champion was Quine. His procedure for determining our ontological commitments was essentially this: identify our best scientific theories; regiment them in the language of first-order logic; then read off the commitments from the resulting regimentations, by seeing what the first-order variables range over. Thanks to Quine and others, by the mid-twentieth century first-order logic had become dominant. Although its connection to logics studied by the ancients was at best unclear, and its callowness undeniable, it soon came to be known—in a brilliant PR coup—as 'classical logic'.

The decades since have seen a great proliferation of logics. Some logics have been offered as *extensions* of first-order logic (e.g. various modal logics); some as *replacements* (e.g. relevant logics or intuitionistic logic); and many other logics have been studied for their mathematical interest. With the chief partial exception of intuitionistic logic, however, all these logics have usually been investigated in a classical first-order metalanguage.[3] First-order Zermelo Fraenkel Choice (ZFC) set

---

[1] First-order logic is always understood to include identity, unless otherwise stated.

[2] For the history of this emergence, see especially Moore (1980, 1988) and Shapiro (1991, ch. 7). We use the terms 'one true logic', 'foundational logic', and 'correct logic' (with uniqueness understood in the second and third cases) interchangeably.

[3] As pointed out by many authors, e.g. Williamson (2014). An example of classical reasoning about intuitionistic logic is the usual soundness and completeness proofs for intuitionistic predicate logic with respect to Kripke semantics (see e.g. Troelstra and van Dalen 1988, ch. 2). *Some* metatheoretic results about intuitionistic logic, such as the soundness and completeness of various such logics with respect to Beth semantics, are intuitionistically acceptable (see again Troelstra and van Dalen 1988, ch. 13).

One True Logic: A Monist Manifesto. Owen Griffiths and A.C. Paseau, Oxford University Press.
© Owen Griffiths and A.C. Paseau 2022. DOI: 10.1093/oso/9780198829713.003.0004

theory remains mathematicians' preferred foundational theory.[4] Metamathematical investigations are usually understood to take place within ZFC (or occasionally some first-order strengthening of it, or some close variant of it). The study of rivals to classical logic almost always takes place in such a metatheory. As for extensions of classical first-order logic, such as quantified modal logic, it remains unclear whether these are logics *sensu stricto*. In any event, these too are typically studied within a metatheory employing classical first-order logic.

So first-order logic still reigns. But as we shall see, it is greatly enfeebled; if in the 1950s it could be likened to the Louis XIV of logics, today it is more like the Louis XVI. Part II's aim will be to hasten first-order logic's demise as *the* foundational logic.

In Part I, we argued for monism, the thesis that there is one true logic. In response to the obvious follow-up—*which* logic is the one true logic?—we have some good news and some bad news. The bad news first: we don't have an exact answer to the question. This book does not contain a systematic argument to the effect that some specific logic is the true one. Lest the reader be unduly disappointed, we hasten to point out that, in our view, no one has come close to answering the question either. We don't know of any convincing arguments that some specific logic is the all-encompassing one true logic.

The good news is that we take some significant steps towards determining the one true logic (OTL). We shall argue that any such logic must be maximally infinitary. Suppose, for instance, that you equate it with first-order logic. Our arguments in Parts II and Part III will show that you should go beyond FOL and adopt at least $\mathcal{L}_{\infty\infty}$ (pronounced 'ell infinity infinity'). This is the logic that extends FOL by allowing quantification over κ variables, for κ any non-zero cardinal (finite or infinite), and allows conjunction or disjunction over κ formulas, again for κ any non-zero cardinal. In $\mathcal{L}_{\infty\infty}$ one can for example formulate a sentence that is true in all and only domains of size $\geq \kappa$ by conjoining κ-many atomic inequations $x_i \neq x_j$ (where i is distinct from j) and then existentially quantifying over κ many of the variables in the resulting conjunction. An informal picture of what such a sentence looks like is:[5]

$$\underbrace{\exists x_1 \exists x_2 \ldots}_{\kappa} (\bigwedge_{1 \leq i < j < \kappa} x_i \neq x_j)$$

Although $\mathcal{L}_{\infty\infty}$ does not have atomic predicates with infinitely many argument places, it can define non-atomic predicates of this kind by means of infinitary

---

[4] *Modulo* the theory's strengthening by large cardinal axioms, virtually all expressible in first-order terms.

[5] This is not entirely accurate, since as we shall see below, on our account sentences of $\mathcal{L}_{\infty\infty}$ have finitely many quantifiers quantifying over sets of variables. But it should help build up the intuitions of anyone unacquainted with infinitary logic.

conjunction or disjunction. Our hypothesis, then, formulated very roughly at this stage, is that the one true logic is as infinitary as can be.

To sharpen Part II's main claim, we must distinguish various dimensions of a logic. This is the task of §4.2.

4.2 Type and sort

Finite relational types are defined as follows:

- i is a type, the type of individuals;
- if $\tau_1, \ldots, \tau_n$ are types, so is the ordered n-tuple $\tau = (\tau_1, \ldots, \tau_n)$, the type of n-ary relations over objects of respective types $\tau_1, \ldots, \tau_n$;
- nothing else is a type.

As an example, take the finite-type hierarchy over the domain of individuals that consists of all people currently alive.[6] Unary relations (properties) over these individuals are of type (i), for example the property *being French*. Binary relations over these individuals are of type (i, i), for example the binary relation *loves*. Unary relations of type $((i))$ are properties of properties of individuals; one such is *being a property TikTok influencers have*, which is instantiated by the property *being an exhibitionist*, of type (i). And so on. Relations in extension may be considered as sets of tuples; for instance, the extension of the binary relation *loves* consists of all and only the ordered pairs of people $\langle a, b \rangle$ such that a loves b. If R is a relation of type $\tau = (\tau_1, \ldots, \tau_n)$ and each a_i is of type τ_i ($1 \leq i \leq n$) then $R(a_1, \ldots, a_n)$ is a statement. (In addition to the above types, we could introduce a new type for statements, but we will not do so here.)

Another example is the finite-type hierarchy over the natural numbers, that is, the hierarchy in which the natural numbers play the role of individuals. In this hierarchy, the property *being prime* is of type (i), *is greater than* of type (i, i), the addition function may be identified with the ternary relation of type (i, i, i), whose extension contains the triple $\langle a, b, c \rangle$ iff $a + b = c$, and so on. As this example shows, functions may be represented as functional relations one arity up.

The type hierarchy classifies relations and individuals. It thereby induces a classification of expressions for each of these items. Names of individuals may in this way be assigned type i and a predicate constant denoting a relation of type τ may itself be assigned type τ.

Which types is FOL concerned with? Models of FOL specify a domain and assign sets of ordered n-tuples to n-adic predicate constants, including sets of ordered

---

[6] For brevity, we often write 'finite-type hierarchy' instead of 'finite-relational-type hierarchy'.

pairs to the identity relation. Its types thus include the type i of individuals as well as all types which in our notation require a single pair of parentheses: (i), (i, i), (i, i, i), and so on. In addition, FOL contains Boolean operations on statements, to be discussed in Chapter 6, and it also includes the existential and universal quantifier, each of type $((i))$. The existential quantifier's interpretation over any domain, for instance, is the relation *being an instantiated property of type (i)*. FOL treats individuals differently from other types, as its language has variables, which may then be quantified over, for individuals—these are the first-order variables. It has predicate constants that denote relations over these individuals, but no variables for the latter (and *a fortiori* no means of quantifying over them), nor any other kind of variables.

Second-order logic (SOL) is concerned with exactly the same range of types as FOL.[7] It differs from FOL by allowing variables (and subsequent quantification over them) in its language over the types (i), (i, i), (i, i, i), etc. Third-order logic, depending on exactly how it is understood, concerns the above types as well as types such as $((i))$, $((i), (i))$ and so on. As the example of FOL and SOL illustrates, some logics may admit constants for the same types but variables for different types.[8]

The logic of Russell and Whitehead's *Principia Mathematica*, in contrast to FOL and SOL, is concerned with *all* types in the relational type hierarchy (and admits variables and quantification over all of them). In the *Simple Theory of Types*, a great simplification of *Principia*'s system,[9] entities are built 'linearly' on the type of individuals. Individuals are of type i, entities of type (i) consist of individuals, the next type of entities is $((i))$, and so on for an arbitrary finite number of steps. Later in this chapter, we will present a logic loosely inspired by that of *Principia* that we call FTT ('finite type theory').[10]

Logics concerned with types in the relational type hierarchy may thus be naturally classified along two dimensions.[11] The first dimension is the type of entities the logic is concerned with; the second are the operations it admits. Taking the first dimension first: as observed, FOL is a logic of individuals and properties/relations over those individuals.[12] It admits individual constants, indi-

---

[7] Throughout this book, we assume standard or full set-theoretic semantics for SOL, in which for example monadic second-order variables range over all subsets of the domain.

[8] In addition, one can formulate logics which have variables of a certain type but no constant letters of this type.

[9] Owed to Ramsey (1925).

[10] There are *many* differences between this logic and the actual logic of *Principia*, perhaps the main one being that FTT does not stratify types into orders.

[11] Among many others, of course. To state the obvious, the fact that objects in a two-dimensional plane may be classified by their x and y coordinates does not preclude them from being classified in other ways. Moreover, as we shall shortly see, the word 'dimension' here should not be taken to imply a strictly linear ordering.

[12] Henceforth we generally use 'relations' to mean 'property and relations', a property being understood as a special type of relation, viz. a unary relation.

vidual variables which may be quantified, and predicate constants, although not predicate variables (so *a fortiori* it disallows predicate quantification). More generally, a logic's first 'dimension' is determined by the types of entities its language represents, and whether for each such type, it allows constants and/or variables, and, if the latter, whether it allows quantification over these variables. This dimension might be called the logic's *type*. Thus FOL's type is specified by saying that it has constants and variables which may be quantified over for individuals, and constants for all relations of individuals; we may write FOL's type as τ_{FOL}. SOL has a different type, as explained earlier, and as we shall see, so does FTT.

The second dimension of a logic in the relational type hierarchy builds on the first. This dimension is determined by the kind of operations the logic allows on formulas of its language. In a second coinage, we call this dimension of a logic its *sort*.[13] FOL allows finite Boolean operations, so that for example if ϕ and ψ are formulas so are $\phi \wedge \psi$, $\phi \vee \psi$, $\phi \leftrightarrow \psi$, $\neg\phi$, $\neg\psi$, and so on; and it allows finite blocks of quantifiers (universal or existential), so that if ϕ is a formula, so is $Q_1 x_1 \dots Q_n x_n \phi$, where each Q_i is either $\forall$ or $\exists$. SOL's sort is identical to FOL's: it too allows finite blocks of quantifiers (first-order and second-order in this case) and finite Boolean operations on formulas.

Infinitary logics typically differ from both FOL and SOL along this dimension. Some readers might be familiar with $\mathcal{L}_{\omega_1\omega}$, which extends FOL by allowing conjunctions and disjunctions of countably infinite length (of well-formed formulas). In contrast, $\mathcal{L}_{\omega\omega_1}$ extends FOL by allowing quantification over a countable infinity of variables (that range over individuals).[14] More generally, when κ and λ are infinite cardinals, the logic $\mathcal{L}_{\kappa\lambda}$ extends FOL by allowing conjunction or disjunction over $< \kappa$ well-formed formulas, and existential quantification over $< \lambda$ variables. In this terminology, FOL is $\mathcal{L}_{\omega\omega}$ or FOL$_{\omega\omega}$ as we may also write it; in other words, it's the logic with FOL's type (τ_{FOL}) and FOL's sort ($\omega\omega$). $\mathcal{L}_{\infty\lambda}$ extends FOL by allowing us to conjoin or disjoin κ well-formed formulas for any cardinal κ and to quantify over $< \lambda$ variables. Symmetrically, $\mathcal{L}_{\kappa\infty}$ extends FOL by allowing us to conjoin or disjoin $< \kappa$ well-formed formulas and to quantify over λ variables for any cardinal λ. Finally, the logic $\mathcal{L}_{\infty\infty}$ allows conjunction or disjunction over κ well-formed formulas and quantification over λ variables for any cardinals κ and λ. In perhaps more transparent notation, $\mathcal{L}_{\kappa\infty}$, $\mathcal{L}_{\infty\lambda}$, and $\mathcal{L}_{\infty\infty}$ could be called FOL$_{\kappa\infty}$, FOL$_{\infty\lambda}$, and FOL$_{\infty\infty}$ respectively.

The infinitary logics $\mathcal{L}_{\kappa\lambda}$, $\mathcal{L}_{\infty\lambda}$, $\mathcal{L}_{\kappa\infty}$ (one per cardinal κ and cardinal λ),[15] and their extension $\mathcal{L}_{\infty\infty}$, all extend FOL by adding operations of greater strength to

---

[13] For the avoidance of misunderstanding, this use of 'sort' bears no relation to many-sorted logic, as discussed by Quine and others, a recent treatment being Barrett and Halvorson (2017).

[14] We use the conventional notation for these logics, and do not distinguish between the ordinal ω and the cardinal $\aleph_0$, or ω_1 and $\aleph_1$, etc.

[15] Where κ and λ are infinite cardinals—this is understood whenever appropriate.

it, without the addition of any further types. In our terminology, they extend FOL along the second dimension, but not the first: they extend its sort but not its type, with the variables quantified over remaining first-order. If precision on this score is wanted, $\mathcal{L}_{\kappa\lambda}$, $\mathcal{L}_{\infty\lambda}$, $\mathcal{L}_{\kappa\infty}$, and $\mathcal{L}_{\infty\infty}$ may be written as $\mathcal{L}_{\kappa\lambda}^{\tau_{FOL}}$, $\mathcal{L}_{\infty\lambda}^{\tau_{FOL}}$, $\mathcal{L}_{\kappa\infty}^{\tau_{FOL}}$, and $\mathcal{L}_{\infty\infty}^{\tau_{FOL}}$; the superscript marks the first dimension and the subscript the second. In this terminology, FOL could be written as $\mathcal{L}_{\omega\omega}^{\tau_{FOL}}$.

Suppose we start off with SOL instead of FOL. We can build a similar hierarchy of infinitary logics on its basis. The logic $SOL_{\kappa\lambda}$ extends SOL ($= SOL_{\omega\omega}$) by allowing conjunction or disjunction over $< \kappa$ well-formed formulas, and quantification over $< \lambda$ predicate or individual quantifiers. Similarly, the logics $SOL_{\infty\lambda}$, $SOL_{\kappa\infty}$ and $SOL_{\infty\infty}$ are the respective SOL-analogues of $FOL_{\infty\lambda}$, $FOL_{\kappa\infty}$, and $FOL_{\infty\infty}$. In particular, $SOL_{\infty\infty}$ extends SOL by allowing conjunctions and disjunctions of any number of formulas, and quantification over any number of variables. As above, $SOL_{\kappa\lambda}$, $SOL_{\infty\lambda}$, $SOL_{\kappa\infty}$, and $SOL_{\infty\infty}$ may be respectively written as $\mathcal{L}_{\kappa\lambda}^{\tau_{SOL}}$, $\mathcal{L}_{\infty\lambda}^{\tau_{SOL}}$, $\mathcal{L}_{\kappa\infty}^{\tau_{SOL}}$, and $\mathcal{L}_{\infty\infty}^{\tau_{SOL}}$.

As we have seen, $FOL_{\infty\infty}$ and SOL are respective sort and type extensions of FOL. Each of these is in turn extended by $SOL_{\infty\infty}$. Here is a picture of these extension facts:

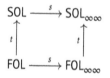

Horizontal arrows, labelled 's', denote sort extensions; vertical arrows, labelled 't', type extensions. Naturally, several logics of intermediate strength lie along the square's edge; the diagram is *not* intended to be a complete representation of the logics from FOL to $SOL_{\infty\infty}$, far from it. Equally clearly, we hope, type and sort are not supposed to be linear dimensions. By this we mean that FOL, for instance, has sort extensions that do not extend one another properly or improperly, such as $\mathcal{L}_{\omega_1\omega}$ (aka $FOL_{\omega_1\omega}$) and $\mathcal{L}_{\omega\omega_1}$ (aka $FOL_{\omega\omega_1}$). The logic FTT extends SOL and FOL to all finite types and its sort extension $FTT_{\infty\infty}$ extends $SOL_{\infty\infty}$ and $FOL_{\infty\infty}$. A more formal discussion of FTT and $FTT_{\infty\infty}$ awaits in §§4.6–4.7.

Finer-grained classifications of sorts are also available. The astute reader will have noticed that when we introduced $SOL_{\kappa\lambda}$, we could have introduced a further distinction between infinitary extensions of SOL. We could have distinguished those that allow quantification over a certain number of first-order variables from those that allow quantification over a different number of second-order variables; and within the latter, we could have also distinguished those that allow quantification over different numbers of second-order variables of different adicities. The logic $SOL_{\kappa_1\kappa_2\kappa_3}$ (note the *three* indices), for instance, is the extension of second-order logic in which $< \kappa_1$ formulas may be conjoined or disjoined, universal or existential quantification over $< \kappa_2$ *first-order* variables is allowed, and universal or existential quantification over $< \kappa_3$ *second-order* variables is allowed. Similarly for $SOL_{\infty\kappa_2\kappa_3}$, $SOL_{\kappa_1\infty\kappa_3}$, $SOL_{\kappa_1\kappa_2\infty}$, $SOL_{\infty\infty\kappa_3}$, $SOL_{\infty\kappa_2\infty}$, $SOL_{\kappa_1\infty\infty}$, and $SOL_{\infty\infty\infty}$. As our justification for infinitary logic will take us all the way to logics of the maximal sort, we will have no use for such fine distinctions. So we continue to write $SOL_{\infty\infty}$, in which expression the second index rolls up into one the number of first-order and second-order variables one may quantify over. Although our focus here is on classical logic (see the Prologue), one could also consider sort-extensions of some of the other logics mentioned in Part I, for example relevance logic.

4.3 The L∞G∞S Hypothesis

We've seen that we can classify logics in the finite-type hierarchy by their type and sort. Our main thesis in Part II is that any such candidate for OTL must have maximally infinitary sort, that is, $\infty\infty$. Or to put it another way, that if the one true logic $\mathcal{L}$ is among these and has type τ then it is identical to $\mathcal{L}^{\tau}_{\infty\infty}$ (the superscript indicating the logic's type and the subscript its sort). The one true logic's formulas are thus closed under κ-ary conjunction and disjunction, and under κ-ary quantification (meaning quantification over κ-many variables) over all its variable types, for all cardinals κ (finite and infinite).

As we had occasion to mention at the start of this chapter, branding matters. With this in mind, we call the *L∞G∞S Hypothesis* the claim in this section's opening paragraph, namely

If OTL is a sublogic of $FTT_{\infty\infty}$ then its sort is $\infty\infty$.

It's worth pausing to explain why we think the label, though slightly tongue-in-cheek, is apt. The English word 'logic' is derived from the Greek 'λόγος'. Over its long history, the latter's meaning has spanned many distinct but related ideas, including a word or sentence or more generally what is said, what is thought, as well as an explanation, reason, or account. In its specifically Christian use in John's Gospel, it meant something akin to an ordering principle. This is the sense John had in mind when he started off his narration by saying that in the beginning was the Word. What we are defending in the logical context—*monism* (or *monologicism*)—is the analogue of monotheism in the religious one. 'L∞G∞S' combines in a few symbols the idea that the one true logic's sort is $\infty\infty$; and by verbal association with John's use of its Greek etymon, it underlines the one true logic's univocality, fundamentality, and comprehensiveness. It's a fitting label for our claim, although we urge our readers to join us in not taking it *too* solemnly.

The L∞G∞S Hypothesis is a precise claim about all logics in the finite-type hierarchy; alternatively, for any sublogic of $FTT_{\infty\infty}$. As such, it does not take in all candidates for the one true logic. Many logics have a type that can be easily described by reference to the hierarchy of finite relational types. But not all do. An example is a logic that admits infinitary types, extending the finite type hierarchy into the transfinite, whose type transcends anything describable by means of finite relational types. Another example is modal logic, not easily shoehorned into the finite relational-type hierarchy either.

Such examples and others do *not* show that logics cannot be decomposed along the two dimensions of type and sort. What they show rather is that the finite-type framework for describing a logic's type applies to many standard logics but not all. A more capacious framework would take in other logics too. Since our interest is not directly in classifying all logics—we find such attempts procrustean—we shall not propose a framework of this kind. Nevertheless, our hypothesis may be strengthened in two ways, one precise, the other (inevitably) more vague.

The first strengthening: the maximal sublogic of OTL that is also a sublogic of $FTT_{\infty\infty}$ is of sort $\infty\infty$. Less formally: the 'intersection' of OTL and $FTT_{\infty\infty}$ is of sort $\infty\infty$. If, for example, OTL is 'second-order logic plus modal operators', it must include $SOL_{\infty\infty}$ as a sublogic. This strengthening maintains the interest of the LꝏGꝏS Hypothesis should it turn out that OTL is not a sublogic of $FTT_{\infty\infty}$. It will generally be clear enough how to extend our arguments for the baseline LꝏGꝏS Hypothesis to support this strengthening as well. We will therefore elide the difference between the two, as we did in the previous-to-last sentence.

An even stronger form of the LꝏGꝏS Hypothesis would be to suppose that the true logic's sort is maximally infinitary, whatever its type might be (i.e. whether or not its types are a subset of the finite relational ones). How to spell out this stronger hypothesis depends of course on how exactly to generalize the notion of sort to all logics whatsoever. The idea would be that the one true logic stands to the sort-weakest logic with which it shares its type as $FOL_{\infty\infty}$ ($= \mathcal{L}_{\infty\infty}$) stands to $FOL_{\omega\omega}$ ($= FOL$). Or in equational terms, that if $\mathcal{L}$ is the one true logic then $\mathcal{L} = \mathcal{L}_{\infty\infty}$. For example, if your preferred foundational logic is first-order logic extended with a handful of sentential operators, then you should take the one true logic to be the extension of this logic closed under the conjunction and disjunction of any number of formulas and quantification over any number of variables (i.e. κ-many for any infinite cardinal κ) and closed under the iteration of as many operators as possible. More generally still, whatever your preferred foundational logic, its foundational operations (Boolean operations, quantification, etc.) should be applicable to any number of formulas or terms/variables or whatever other iterable expressions the logic's language contains. The LꝏGꝏS Hypothesis generalizes in this way to all logics whatsoever.[16]

---

[16] There is a still further generalization available here. Sublogics of $FTT_{\infty\infty}$ (including all sublogics of $\mathcal{L}_{\infty\infty}$, i.e. $FOL_{\infty\infty}$), all have finite alternations of quantifiers. As we shall see in more detail shortly, the inductive definition of $\mathcal{L}_{\kappa\lambda}$ means that existential quantifiers can only alternate finitely many

From the perspective of contemporary philosophical, mathematical, and logical practice, the L∞G∞S Hypothesis is already radical enough. With a few exceptions, notably Alfred Tarski and Gila Sher, philosophers or logicians have not seriously contemplated, still less have they espoused, the idea that our fundamental logic—the logic that captures the correct consequence relation—is infinitary. In this climate, our hypothesis demands spelling out and defence, before we can even start to think about OTL's type. Arguing for a highly infinitary logic is sufficiently radical that it will take up the best part of this book. In §4.4, we briefly tell the story of infinitary logic's birth, fall, and recent resurgence.

4.4 Back to origins

Although radical in the present intellectual climate and downright heretical in that of the mid-twentieth century, our proposal that logic is infinitary would have found much more sympathy among contemporary logic's pioneers. Modern logic took off in the late nineteenth century. Its pioneers, such as Frege, Russell and Whitehead, Hilbert, Pierce, Schröder, and many others worked in what we would now consider a higher-order logic. The theory of types tended to be the standard logical system in the 1920s. At some point in the 1930s, FOL, originally considered by Hilbert and many others as a partial subsystem of all of logic—that is, the theory of all finite types—emerged as the canonical dominant logical system.[17]

A less familiar fact is that several of these pioneers made use of infinitary formulas.[18] Writers such as Eugen Müller (1910), Hilbert (1905), Löwenheim (1915), Skolem (1920), and, very self-consciously so, Zermelo (1932), all deployed infinitary formulas.[19] It is not until Skolem (1923) that we find a fairly clear claim to the effect that finitary first-order logic is all of logic, following Skolem's (1922) clarification of Zermelo's notion of a 'definite proposition' as one that can be expressed in first-order terms. Yet only two or three years prior to that, Skolem himself had happily used infinitely long strings of quantifiers as well as infinite conjunctions and disjunctions.[20]

---

times with universal quantifiers. (If the only primitives are existential quantifiers then a universal quantifier should be understood as an existential quantifier preceded and succeeded by negation). An even more infinitary hypothesis, which we leave for further work, would be to consider logics with infinite alternations of quantifiers.

[17] Moore (1990, pp. 47–8). The history of higher-order logic's rise and fall in the 1930s has been carefully traced in the literature: see in particular Moore (1988) and Eklund (1996).

[18] Moore (1990, 1997) details the history.

[19] See Moore (1990) for relevant citations. Moore (1997) lists several other authors. Both of Moore's articles discuss ambiguous cases such as Schröder.

[20] The cases of Wittgenstein and Ramsey are less clear. Ramsey cites Wittgenstein approvingly as saying that there is no reason that a truth-function should not have infinitely many arguments, but uses quantification notation rather than introducing any sort of infinitary syntax (Ramsey 1925, p. 343). So as Moore (1990, p. 53) points out, there is a sense in which Ramsey did not actually countenance

By the mid-twentieth century, then, as a result of these and other influences, FOL had emerged as most logicians' foundational logic of choice. And where the immediate successors of the late nineteenth/early twentieth-century trailblazers went, there the next generation of logicians and their sequacious students followed. As a consequence, higher-order and infinitary logics were relatively neglected. FOL's emergence went hand in hand with the submergence of higher-order logic and infinitary logic. These were two sides of the same coin: if FOL is the one true logic then the study of stronger logics becomes a piece of relatively recondite mathematics, of little obvious philosophical interest. Naturally, the emergence of one logic need not lead to the submergence of all others. Propositional logic, which was isolated from stronger systems and studied in its own right, is a case in point.[21] But propositional logic is a fragment of FOL, so its study may be seen as the study of an aspect of the one true logic itself.

Higher-order quantification's fall from grace has been partially reversed in recent decades. Its advocacy at the hands of George Boolos, and later philosophers who took up his mantle, has done much to restore SOL's philosophical legitimacy.[22] The story of infinitary logics is more mixed. Moore (1990) dates the birth of 'modern infinitary logic' to the mid-1950s with the work of Henkin, Karp, Scott, and Tarski on infinitely long formulas.[23] Since then, much mathematical work on these logics has been carried out, a milestone being the publication of Barwise and Feferman (1985). But the milestone turned out to be something of a capstone, for by the mid-1980s work on infinitary logic had run out of steam and mathematical logicians' attention refocused on first-order logic and first-order theories.

On the philosophical side, the Tarski–Sher Thesis has picked up some momentum since the publication of Gila Sher's 1991 book *The Bounds of Logic* and

---

infinitary logic. *Pace* Moore (1990, p. 49), the same might be said about Peirce, who claims that existential and universal quantification over any domain, even an infinite one, are *similar* to forming an infinite sum or product, but does not go as far as to say that they are *identical*. The case of Wittgenstein is more complicated. Moore (1990, p. 52) claims that infinitely long formulas arose in the *Tractatus*. He appears to have the N-operator in mind (see the 5.5s and 6–6.021 of the *Tractatus* for relevant passages), which applies to a plurality of propositions and yields their joint negation. Yet at 4.2211 in the *Tractatus*, perhaps the clearest indication of his thinking here, Wittgenstein says no more than that *there could be* infinitely many facts. At most, then, he leaves it an open question whether N could apply to infinitely many propositions. As Wittgenstein's exposition of these matters is less than lucid, we leave it to Wittgenstein scholars to interpret him better. At present, we feel unable to include him as a past partisan of the infinitary cause.

[21] See Post (1921), which carves the propositional calculus out of the system of *Principia Mathematica*.

[22] See Tharp (1975) and Resnik (1988) for influential attacks on second-order logic. Shapiro (1991) offered an extensive defence of second-order logic, based in large part on its suitability for mathematical practice. More recently, Trueman (2012) has offered a defence of second-order logic, arguing that its ontological commitments are problematic only if those of first-order logic already are.

[23] Henkin and Tarski organized a seminar on infinitary logic at Berkeley in the autumn of 1956, at which Karp reported some of the results in her thesis (Moore 1990, p. 56). Karp's supervisor was Henkin, whose earliest work on infinitary logic dates from 1954 (Moore 1997, pp. 106–7).

later work by Sher and others.[24] But it's fair to say that, among philosophers at least, infinitary logic is still largely seen as a mathematical curiosity. To be sure, infinitary logics are of mathematical interest; but they are not usually taken to have foundational significance. To return to the distinction made in the Prologue, the study of infinitary logics is regarded as a part of pure rather than applied logic. And this study, like all other metatheoretic investigations, takes place in a finitary metalogic.

This book is, in contrast, intended as a contribution to *applied* logic. We join Sher and a handful of others in the fight against the finitarist usurpation of logic. As this small but growing band sees it, the modern pioneers were right not to restrict the range of logic to FOL, or any other finitary logic. What the pioneers lacked, however, was a principled reason for doing so. Or better: although the motivation in terms of generality and topic-neutrality was available to them, they lacked the mathematical tools to develop it in the right way. If we are right, the era of finitary logic's foundational primacy ought to be relatively short-lived; and if we are listened to, it will be. There is just as much reason to go beyond FOL's sort as there is to go beyond its type. Infinitary logic should be restored to grace in the same way that higher-order logic partially has. Infinitary logic is not merely of considerable mathematical interest; it also offers the right account of logical consequence.

4.5 Top-down and bottom-up arguments

We now summarize how Part II will help achieve our goal of arguing for our desired conclusion, *viz.* the LᴏᴏGᴏᴏS Hypothesis. Suppose we wish to draw a dichotomy in a particular domain. A top-down approach would be to characterize it theoretically. Everything on one side of the divide has some property, whereas everything on the other side lacks it. In the right circumstances, such a characterization might enable us to classify particular cases as falling on one side or other of the divide. A successful top-down approach must not only draw a line across a particular domain; it must also explain *why* the line is drawn precisely there. In the case of logical consequence, the domain consists of natural-language arguments,[25] and the task is to distinguish valid arguments from invalid ones. For the top-down approach to have any chance of success, it must draw on notions central to our conception of logical consequence, or at least notions central to the way of thinking about logical consequence in the tradition we take ourselves to be part of.

---

[24] For later work by Sher, see especially her (2001), (2013), (2016) and, for the most recent overview, (2021).

[25] More precisely, in the terminology of the Prologue, E_c^+-arguments (arguments of cleaned-up and extended English).

The bottom-up approach, in contrast, starts with particular cases. It appeals to existing practice to classify individual cases on one side of the divide or the other. It may also help itself to relatively uncontentious principles emerging from this classification. Proceeding in this way, we extend the range of cases we can confidently classify, letting extension and counter-extension grow together. Sufficient proximity to paradigm cases allows us to classify a particular example on the positive side; sufficient proximity to foils allows us to place it on the other. The process here is more generally abductive.[26] At its most powerful, the bottom-up approach provides enough such instances that classification of *all* cases can proceed by comparison with them. In the case of logical consequence, in which the domain consists of natural-language arguments, we start with clear-cut cases of validity and invalidity. The bottom-up approach extends the consequence relation's extension and counter-extension in this piecemeal way, or by using relatively light theoretical machinery.

Whenever available, the ideal approach is to combine *both* top-down and bottom-up approaches. And fortunately, this combination *is* available here and will be implemented in Part II. Chapter 5 will provide a host of examples that suggest that FOL and SOL's sort is too limited to account for natural-language validity. This bottom-up argument will suggest that the one true logic's sort ought to be $\infty\infty$. The inadequacy of FOL and of SOL's sort is revealed in this bottom-up way. The reason we focus on FOL and SOL in Chapter 5 is because of their current popularity as candidates for the one true logic.

Chapter 6 is a top-down argument to the same effect. We ground logicality in topic-neutrality. This vindicates a condition on logical relations and constants that supports a logic of maximally infinitary sort. Chapter 7 concludes our presentation of what we know and what we don't know about the one true logic. Part III will then be concerned with objections to the picture presented in Part II.

If we are right, then, the one true logic has maximally infinitary sort. The rest of this chapter is devoted to filling in some of the formal background, and in particular to specifying the logic $FTT_{\infty\infty}$, and thereby its sublogics as well.

4.6 Pure $FTT_{\infty\infty}$ defined

Our focus in what follows will be on expressions denoting an element of a finite type hierarchy over a domain of individuals. This covers a good deal of language, including:

---

[26] And should not be equated with a naive inductivism in which any counterexample disconfirms the theory.

- predicates, which denote properties and relations;
- names of functions, which may be cast as relations in the usual way:

  $f(x_1, \ldots, x_n) = x_{n+1}$ iff $R_f(x_1, \ldots, x_n, x_{n+1})$;
- names, which denote individuals or properties and relations ('Aristotle' is an example of the former and 'wisdom' of the latter);
- quantifiers, which may be cast as higher-order properties or relations in a standard way.

The logic $\mathcal{L}_{\infty\infty}$ extends FOL by allowing the conjoining or disjoining of κ-many well-formed formulas for any κ, and by allowing existential or universal quantification over κ-many variables for any infinite κ. Its semantics also extends that of FOL in the obvious way. $\mathcal{L}_{\infty\infty}$ can be thought of as the most infinitary version of first-order logic, and we may write it as FOL$_{\infty\infty}$.

We may generalize $\mathcal{L}_{\infty\infty}$ to higher finite or relational types. The language of $\mathcal{L}_{\infty\infty}^{FTT}$, or FTT$_{\infty\infty}$ as we may alternatively call it, is the language obtained from $\mathcal{L}_{\infty\infty}$ by allowing variables of all finite types, and quantifiers over them. (From here on, 'types' means finite types, unless otherwise specified.) Thus as well as all the $\mathcal{L}_{\infty\infty}$-variables $x_0^i, x_1^i, \ldots x_n^i, \ldots$ which range over the objects (individuals) of a given model, the language of $\mathcal{L}_{\infty\infty}^{FTT}$ contains second-order variables $X_0^{(i)}, X_1^{(i)}, \ldots, X_n^{(i)}, \ldots$ which range over subsets of the domain, third-order variables $X_0^{((i))}, X_1^{((i))}, \ldots, X_n^{((i))}, \ldots$ which range over subsets of subsets of the domain, variables $X_0^{(i,(i))}, X_1^{(i,(i))}, \ldots, X_n^{(i,(i))}, \ldots$ which range over ordered pairs whose first element is an object of the domain (i.e. an individual) and whose second element is a subset of the domain, and so on. The logic $\mathcal{L}_{\infty\infty}^{FTT}$ thus allows quantification over any number of relations of any finite type, as well as Boolean operations on any number of formulas. From now on, we call this logic FTT$_{\infty\infty}$.

With the informal picture in hand, let's now define *pure* FTT$_{\infty\infty}$ precisely (sometimes the epithet 'pure-equality' is used instead of 'pure'). This logic is 'pure' because it lacks non-logical vocabulary; the pure fragment of first-order logic in this same sense consists of first-order logic with no non-logical symbols.[27] For example, $\exists x(x = x)$ is a sentence of both pure FTT$_{\infty\infty}$ and pure FOL, whereas $\exists x Fx$ is a sentence of both FTT$_{\infty\infty}$ and FOL, but not a sentence of their pure versions, as it contains the non-logical symbol F.

It is worth noting that our use of 'logical' and 'non-logical' in the previous paragraph is simply a means of distinguishing one type of vocabulary item from another. We have availed ourselves of this standard terminology to promote readability and avoid fussiness. Naturally, the question remains whether 'logical' items are really logical, a question which the rest of Part II will go on to answer, in the affirmative. Our use of 'logical' and 'non-logical' here as sorting devices does

---

[27] Similarly for second-order logic. A different way of understanding 'pure second-order logic' as a logic containing second-order but no first-order quantifiers is discussed in Paseau (2010). A yet third way is that of Church, who takes a calculus to be pure if it has no constants.

not prejudge that question. In any case, as we are about to see, it is easy enough here to specify the 'logical' vocabulary in extensional terms simply by listing it.

The vocabulary of $FTT_{\infty\infty}$ includes, for any cardinal κ and type τ, κ-many variables $X_0^\tau, X_1^\tau, \ldots$ of type τ. It also includes the identity symbol $=$; the negation symbol $\neg$; the disjunction symbol $\bigvee$; the existential quantifier $\exists$; the comma , ; left parenthesis (; and right parenthesis).

Any formula of the form $X_i^\tau = X_j^\tau$, where X_i^τ and X_j^τ are both (distinct or identical) variables of type τ is atomic. Any formula of the form $X^{(\tau_1, \ldots, \tau_n)}(X^{\tau_1}, \ldots, X^{\tau_n})$, where $\tau_1, \ldots, \tau_n$ are types, is also atomic.

Atomic formulas are all formulas. If ϕ is a formula, so is $\neg\phi$. The interpretation of $\neg$ is fixed as negation, so that the formula $\neg\phi$ will be true in a model under an assignment iff ϕ isn't. (The semantics will follow in §4.7; we are merely foreshadowing it here.) If $\{X_i : i < \kappa\}$ is a set of κ-many variables (of any type) and ϕ is a formula, then $\exists\{X_i : i < \kappa\}\phi$ is a formula. In the semantics, the interpretation of $\exists\{X_i : i < \kappa\}$ will be existential quantification over all the variables in the set $\{X_i : i < \kappa\}$. If for $i < \kappa$ each of the ϕ_i is a formula, then $\bigvee\{\phi_i : i < \kappa\}$ is also a formula. The interpretation of $\bigvee$ is fixed as disjunction, as we shall shortly see. We may also include $\bigwedge$ as a primitive vocabulary item, or use it as shorthand for $\neg\bigvee\neg$ (i.e. $\bigwedge\{\phi_i : i < \kappa\}$ abbreviates $\neg\bigvee\{\neg\phi_i : i < \kappa\}$). Nothing is a vocabulary item of the pure language of $FTT_{\infty\infty}$, or is one of its atomic formulas, or is one of its formulas, unless it is required to be by the rules in this and previous paragraphs.[28]

The impure language of $FTT_{\infty\infty}$ also contains non-logical constants. It contains non-logical constants of each type, for example the constant a for individuals. It more generally contains non-logical constants of the form R^τ of type τ. (In this notation, an individual constant is R^i, where i is the type of individuals.) It also contains non-logical function symbols, though these do not add logical power since a function whose arguments are of type $\tau_1, \ldots, \tau_n$ and value of type τ may be represented as a functional relation of type $(\tau_1, \ldots, \tau_n, \tau)$. One can add κ-many of all these vocabulary items for any cardinal κ. The language and grammar of full $\mathcal{L}_{\infty\infty}^{FTT}$ are a straightforward extension of those of the pure fragment, as is its semantics (see §4.7).

Given this general characterization, specifying a particular sublogic of $FTT_{\infty\infty}$ is straightforward. Take for example the much-studied logic $\mathcal{L}_{\omega_1\omega} = FOL_{\omega_1\omega}$.[29] Let FOL^* be first-order logic with a countable infinity of relation, function, and constant symbols, ω_1-many variables, as well as parentheses and the usual logical vocabulary. The latter includes a countable and truth-functionally complete set of connectives containing *inter alia* the negation symbol $\neg$, the conjunction symbol $\wedge$ and the disjunction symbol $\vee$.[30] The last two symbols are usually written as $\bigwedge$ and $\bigvee$ when they conjoin statements in a set. The syntax of $\mathcal{L}_{\omega_1\omega}$ is then given by:

---

[28] We have not distinguished formulas from preformulas for simplicity.
[29] For much more on $\mathcal{L}_{\omega_1\omega}$, see Keisler (1971).
[30] NB In this book, 'countable' on its own always means 'finite or countably infinite'.

Vocabulary   The vocabulary of $\mathcal{L}_{\omega_1\omega}$ is that of FOL*.

Atomic Formulas   Atomic FOL*-formulas are all and only atomic FOL-formulas.

Grammar   The class of all well-formed formulas of $\mathcal{L}_{\omega_1\omega}$ is the least class $\mathcal{C}$ such that:

(i)   Each atomic FOL*-formula is in $\mathcal{C}$;

(ii)   If ϕ is in $\mathcal{C}$ and v is a variable, then $(\neg\phi)$ is in $\mathcal{C}$, $\forall v\phi$ is in $\mathcal{C}$, and $\exists v\phi$ is in $\mathcal{C}$;

(iii)   If Φ is a countable (finite or countably infinite) non-empty subset of $\mathcal{C}$, then $\bigwedge\Phi$ is in $\mathcal{C}$, and $\bigvee\Phi$ is in $\mathcal{C}$.

We may similarly define any sublogic of FTT$_{\infty\infty}$, including FOL$_{\kappa\lambda}$ (usually written $\mathcal{L}_{\kappa\lambda}$) for any infinite κ and λ, and FOL$_{\infty\infty}$ (usually written $\mathcal{L}_{\infty\infty}$). When κ is a singular cardinal (see the next paragraph), we have to be a little careful about the definition of FOL$_{\kappa\lambda}$, since we can simulate a conjunction of cardinality κ by a conjunction of lesser cardinality. As our interest will be chiefly in logics of sort $\infty\infty$, we set aside caveats and complications of this kind.

Incidentally, this is as good a place as any to define what it is to be a singular or regular cardinal, seeing as we employed the distinction in the previous paragraph and will do so several more times in Part II. The *cofinality* of an ordinal α is the least ordinal β such that there is an increasing sequence of order-type β that is unbounded in α, that is, has an element greater than any ordinal γ smaller than α. (As customary and convenient, although not necessary, we identify cardinals with the least ordinals of a given cardinality.) A cardinal whose cofinality is smaller than itself is *singular*, and a cardinal whose cofinality is equal to itself is *regular*. So for example, $\aleph_1$ (the first uncountable cardinal and the second infinite one) is regular because an increasing sequence of ordinals unbounded in $\aleph_1$ must be uncountable, the reason being that a countable union of countable sets remains countable. In contrast, $\aleph_\omega$ (the ωth infinite cardinal) is singular because its cofinality is ω: the countable sequence $\aleph_0, \aleph_1, \ldots, \aleph_n, \ldots$ (where n is finite) is unbounded in $\aleph_\omega$.

4.7 Semantics for pure FTT$_{\infty\infty}$

For the pure language of FTT$_{\infty\infty}$ a model $\mathcal{M}$ is given by specifying a domain $D_\mathcal{M}$. We shall assume that $D_\mathcal{M}$ is a non-empty domain on which a hierarchy of finite types is founded. Since $D_\mathcal{M}$ is a set of any objects and of any cardinality, our background metatheory is set theory with urelements. The type of individuals, that is, elements of $D_\mathcal{M}$, is i. The set of type $\tau = (\tau_1, \ldots, \tau_n)$ over $D_\mathcal{M}$ is the power set of $E_\mathcal{M}^{\tau_1} \times \ldots \times$

$E_{\mathcal{M}}^{\tau_n}$, where $E_{\mathcal{M}}^{\tau_j}$ is the set of elements of type τ_j (hence the set of individuals $E_{\mathcal{M}}^{\tau_i}$ is identical to $D_{\mathcal{M}}$). Thus the finite-type hierarchy over D consists of the elements of $\bigcup_{\tau} E_{\mathcal{M}}^{\tau} = R_\omega[D_{\mathcal{M}}]$, the set of sets of finite rank over $D_{\mathcal{M}}$. More precisely,

$$R_0[D_{\mathcal{M}}] = D_{\mathcal{M}};$$
$$R_{n+1}[D_{\mathcal{M}}] = \mathbb{P}(R_n[D_{\mathcal{M}}]);$$
$$R_\omega[D_{\mathcal{M}}] = \bigcup\{R_n[D_{\mathcal{M}}] : n \in \omega\}.$$

Elements of $E_{\mathcal{M}}^{\tau}$ will appear in some $R_n[D_{\mathcal{M}}]$, which n precisely depending on τ. We note pre-emptively that the use of set-sized domains $D_{\mathcal{M}}$ will have implications for our Chapter 6 discussion of isomorphism invariance.

A variable assignment over $D_{\mathcal{M}}$ is an assignment of appropriate values to the variables in V: individual variables are assigned elements of $D_{\mathcal{M}}$, monadic second-order ones are assigned elements of $\mathbb{P}(D_{\mathcal{M}})$, and so on. Thus if σ is a variable assignment and X^τ is a variable of type τ, $\sigma(X^\tau)$ is an element of $E_{\mathcal{M}}^{\tau}$. Hence all the values of V-assignments are elements of $R_\omega[D_{\mathcal{M}}]$.

We define satisfaction for the pure language of $\text{FTT}_{\infty\infty}$ as follows:[31]

$(\mathcal{M}, \sigma) \vDash X^{(\tau_1,\ldots,\tau_n)}(X^{\tau_1}, \ldots, X^{\tau_n})$ iff $\langle \sigma(X^{\tau_1}), \ldots, \sigma(X^{\tau_n}) \rangle \in \sigma(X^{(\tau_1,\ldots\tau_n)})$;

$(\mathcal{M}, \sigma) \vDash X_i^\tau \simeq X_j^\tau$ iff $\sigma(X_i^\tau) = \sigma(X_j^\tau)$;

$(\mathcal{M}, \sigma) \vDash \bigvee\{\phi_i : i < \kappa\}$ iff $(\mathcal{M}, \sigma) \vDash \phi_i$ for some ϕ_i in the set $\{\phi_i : i < \kappa\}$;

$(\mathcal{M}, \sigma) \vDash \neg\phi$ iff it is not the case that $(\mathcal{M}, \sigma) \vDash \phi$, which we write as $(\mathcal{M}, \sigma) \nvDash \phi$;

$(\mathcal{M}, \sigma) \vDash \exists\{X_i^{\tau_i} : i < \kappa\}\phi$ iff there is a variable assignment ρ that differs from σ at most over the variables in the set $\{X_i^{\tau_i} : i < \kappa\}$ such that $(\mathcal{M}, \rho) \vDash \phi$.

A full $\mathcal{L}_{\infty\infty}^{FTT}$-interpretation consists not just of a domain, but also of an interpretation of the non-logical constants over that domain, that is, an assignment of an element of $E_{\mathcal{M}}^{\tau}$ to a non-logical constant of type τ. The satisfaction relation for full $\text{FTT}_{\infty\infty}$ is the obvious extension of these clauses to sentences containing non-logical constants.

Since in a model $\mathcal{M}$ (relative to an assignment σ)[32] the variable X^τ is interpreted as an element of $E_{\mathcal{M}}^{\tau}$, the interpretation of such a formula is a relation whose relata are all relations of finite type over $D_{\mathcal{M}}$. Observe that the interpretation of a pure $\text{FTT}_{\infty\infty}$-formula need not be a relation of finite type itself (over $D_{\mathcal{M}}$ understood). For example, the ω-long formula $\bigvee\{X_k^{(i)} \simeq X_k^{(i)} : k \in \omega\}$ is satisfied by any ω-sequence of relations of type (i) in a model, that is, subsets of $D_{\mathcal{M}}$; but it is

---

[31] The formulas in question are assumed to be well-formed. When greater clarity is sought, as it is here, we use $\simeq$ as the sign for object-language identity, and reserve $=$ for metalanguage identity.
[32] This will generally be understood.

not a relation of type $((i))$, since it has infinitely many relata. We call a relation that is an interpretation of a pure FTT$_{\infty\infty}$-formula a *relation of finite-type relations*, since all its relata are finite-type relations. Clearly, all finite-type relations are relations of finite-type relations, but not vice versa. Chapter 6's formal discussion will be concerned with relations of finite-type relations. That concludes our introduction to the language and semantics of FTT$_{\infty\infty}$.

4.8 What are the formulas of pure FTT$_{\infty\infty}$?

We conclude this chapter with a brief ontological discussion, which may be skipped if desired. We outline three possible accounts (not intended as exhaustive) of a logic's sentences.

When thinking about a finitary logic such as FOL or SOL or propositional logic, a natural first approach would be to take its sentences as expression types. From this point of view, the sentence usually written as

$\exists x Fx$,

is just the expression type instantiated by any of its tokens, such as the one presently on your page or screen.

Finitary languages are usually presented in this fashion. The problem with this approach is twofold. To begin with, a little reflection reveals that the sentences of first-order logic cannot really be expression-types. Think for example of languages that use an alphabet non-overlapping with the English alphabet, in which there are no expression-types of the form 'x' or 'F' that make up '$\exists x Fx$'. And even within a single alphabet, it would be odd to privilege a particular choice of symbols; there is no reason to prefer $F_1, \ldots, F_n, \ldots$ or $G_1, \ldots, G_n, \ldots$ as predicate letters. So the FOL-sentence usually written as '$\exists x Fx$' cannot literally be the expression type '$\exists x Fx$'. Moreover, to suppose that there is a more general sense of expression-type that transcends any particular symbolic expression seems to us a contradiction in terms, at least given what we understand expression-types to be. And nor can one take the sentences to be equivalence classes of expression-types, since *any* symbol can conventionally stand in for any notion.

A second reason against the expression-type approach is particularly relevant in the present context. It is at best not clear that the infinitary sequences of expressions required by FTT$_{\infty\infty}$ exist. For example, it is not clear that an expression of large order-type, say $\aleph_\omega$, exists. At any rate, it would be better to avoid such ontological assumptions about expression-types.

The second, *mixed*, approach is to take infinitary languages to consist of both sets and expressions. This approach is sometimes tacitly adopted in presentations of infinitary languages. These presentations almost always help themselves to set-theoretic vocabulary in their specification of the quantification and disjunction or conjunction clauses. For example, the existential clause often takes the form that if U is a set of variables and ϕ is a formula then $\exists U \phi$ is a formula (similarly for the universal quantifier); and the disjunction clause is usually that if Φ is a set of formulas then $\bigvee \Phi$ is also a formula (similarly for conjunction). The background set theory is assumed to be a theory of sets and urelements, according to which sets such as, say, $\{x_0, x_1, \ldots, x_n\}$ exist, where the x_i are expression-types.

An important disadvantage of the mixed approach is the unclarity about the nature of the 'mixed' expressions. Although sets of expressions are unproblematic, what is it to concatenate such a set with a linguistic expression? What, for example, is $\neg\exists\{x\}(x = x)$, or $\exists\{x\}(x = x)$? We have no natural account of what concatenating a linguistic expression with a set comes to. Much of the motivation for the mixed approach is also undermined once we give up the expression-type approach to sentences of finitary logics. If even finitary sentences are not expression-types, why assume that infinitary ones are, at least in part?

The final, *set-theoretic*, approach is to take infinitary languages to consist entirely of sets. Expressions such as $\exists$, $=$, $\neg$ and so on, are coded as sets, and concatenation is given its usual set-theoretic definition; for example the formula '$x = x$' would be coded as the sequence $\langle a, b, a \rangle$ where a is the set coding the variable x and b the symbol '$=$'. That the coding involves some arbitrary choices is inevitable, but innocuous.

Although the third approach does some violence to the usual way of presenting logical languages, it works smoothly and avoids the problems faced by the other two approaches. On this approach, the language of FOL is a sublanguage of that of $\text{FTT}_{\infty\infty}$ not because its FOL-expressions-types are also $\text{FTT}_{\infty\infty}$-expression-types, but because its sentences are sets that are also $\text{FTT}_{\infty\infty}$-sentences. If pressed to make a choice, this is the approach we would adopt. In any case, since in our book the point of a logic is to represent implicational facts, ultimately the nature of its sentences is irrelevant, just as it is irrelevant for the purposes of astronomical modelling what the spheres in an orrery are made of. What matters instead are the implicational relations between the given 'sentences', whatever they might be.

5

Beyond the Finitary

The one true logic must underwrite the validity or invalidity, as the case may be, of arguments not just of cleaned-up English, but of its potential extensions (E_c^+ in the Prologue's terminology). It must respect all implicational facts, that is, facts about which such sentences logically follow from others. As far as deductive inference is concerned, all we can say at the outset is that there may or may not exist a deductive system that captures all and only these implications.[1]

This chapter offers some 'bottom-up' arguments for first-order logic's and second-order logic's inability to serve as this logic, with the 'top-down' argument reserved for Chapter 6. The fundamental failure of both of these logics is that they have finitary sort (i.e. $\omega\omega$ sort). Extended and generalized, our arguments show that far from being finitary, the one true logic's sort must be $\infty\infty$. In other words, the one true logic is as infinitary as can be.

We start with first-order logic and then move on to second-order logic. We call someone who champions first-order logic as the one true logic a *first-orderist*; and one who champions second-order logic a *second-orderist*.

5.1 Generalized quantifiers

Linguists have for decades now pointed out that first-order logic lacks the resources to define many quantifiers that feature prominently in natural languages. Mathematicians have come to the same conclusion for several quantifiers of mathematical interest.[2] First-order logic, for example, cannot define quantifiers such as 'there are infinitely many', 'there are uncountably many', 'most', and many other quantifiers whose truth-conditions depend on cardinality facts. Since it is an example we will redeploy later, let's consider the generalized quantifier 'most' (called a *determiner* by linguists). 'Most Fs are Gs' is here understood as: there are more F-and-Gs than there are F-and-not-Gs (e.g. there are 52 F-and-Gs and 37 F-and-not-Gs, or uncountably many F-and-Gs and countably many F-and-not-Gs). Following the publication of Mostowski (1957) and Barwise and Cooper (1981, appendix C), most of these indefinability

---

[1] Recall from the Prologue the sharp distinction between implication (modelled by semantic consequence) and inference (modelled by deductive consequence).

[2] See Mostowski (1957) for the first set of indefinability results along these lines.

One True Logic: A Monist Manifesto. Owen Griffiths and A.C. Paseau, Oxford University Press.
© Owen Griffiths and A.C. Paseau 2022. DOI: 10.1093/oso/9780198829713.003.0005

results are now routine exercises. Nevertheless, as its generalization will be important later, we sketch the argument for the FOL-undefinability of 'most'. Suppose 'Most Fs are Gs' were first-order definable by the formula $\mu(F, G)$. The set $\Gamma = \mu(F, G) \cup \{\exists_1(F, \neg G), \ldots, \exists_n(F, \neg G), \ldots\}$, where $\exists_n(F, \neg G)$ is the first-order formalization of 'There are at least n Fs that are not Gs', with n finite, is satisfiable by a model with an uncountable domain such that: all elements of the domain are F; a countable infinity of elements are not-Gs; all the remaining elements (i.e. all but the countable infinity of not-Gs) are Gs. Exploiting the downward Löwenheim-Skolem theorem, this model must have a submodel in which there are more F-and-Gs than F-and-not-Gs, there are a countable infinity of F-and-Gs, but also a countable infinity of F-and-not-Gs. Contradiction. It follows that 'Most Fs are Gs' is not definable by any first-order formula. This argument applies whichever first-order language one might try to define 'most' in. In particular, it applies against any first-orderist who maintains that 'most' can be defined in first-order terms within first-order set theory.

Plausibly, however, generalized quantifiers such as 'most' and others *are* logical. This inherent limitation of FOL has not gone unnoticed in the literature. In what follows, we try to go further than the current literature by advancing the sharpest version of this objection to first-orderism.

One reason for dissatisfaction with the first-order framework, popular among linguists, is that the syntactic structure of first-order quantified statements is very different from that of quantified statements in natural language. Barwise and Cooper (1981, pp. 164–5) give a wealth of examples, including the four sentences in the left-hand column of the table below.

English sentence	FOL-formalization
[Harry]$_{\text{NP}}$ [sneezed]$_{\text{VP}}$	Sa
[Some person]$_{\text{NP}}$ [sneezed]$_{\text{VP}}$	$\exists x(Px \wedge Sx)$
[Every man]$_{\text{NP}}$ [sneezed]$_{\text{VP}}$	$\forall x(Px \rightarrow Sx)$
[Most babies]$_{\text{NP}}$ [sneeze]$_{\text{VP}}$	None

Barwise and Cooper argue that phrases labelled as NPs (Noun Phrases) above belong to a single syntactic category. As they observe, such phrases occur as the subjects of intransitive verbs in the given examples; but the phrases may also appear as the objects of transitive verbs and as objects of prepositions. Indeed, the syntactic mismatch between English sentences and their standard first-order formalizations should be manifest from the table: there is no conjunction in sight in the sentence 'Some person sneezed', just as no conditional is visible in the sentence 'Every man sneezed'.

Natural-language semantics is a project that tries to cleave closely to the syntactic structure of natural-language sentences. (This structure tends to be *theoretical*, delivered by syntactic theory, rather than surface structure.) From this point of view, Barwise and Cooper's criticism, echoed many times since in the linguistics literature, is entirely understandable. From the point of view of a logician whose primary interest is capturing natural language's implicational structure, however, the criticism is less troubling. As we saw in Chapter 2, syntactic constraints on formalization—what we called the grammatical criterion—are of secondary importance. Although they have a role to play,[3] they are generally auxiliary to the goal of reflecting implicational relations. This applies to both surface and theoretical syntax, and it explains why first-orderists are not worried that the syntax of first-order formalizations often bears an oblique relationship to that of the sentences they formalize.

What *should* worry first-orderists is the fact that generalized quantifiers such as 'Most' cannot be represented in first-order terms, as indicated in our table's bottom right-hand corner. For it is very natural to see arguments such as

Most As are Bs
All Bs are Cs
∴ Most As are Cs

as valid in virtue of their logical form. The only argument we have come across or can come up with ourselves for the logicality of 'Most' is a version of our 'top-down' argument. This argument is based on the idea that logic should be topic-neutral and maximally general, about which much more in the next chapter.[4] As such, it is 'top-down' rather than 'bottom-up'.

Since this chapter's arguments are intended to proceed from the bottom up, we consider other sorts of arguments for the logicality of related generalized quantifiers. We focus in particular on cardinality quantifiers such as 'there are infinitely many' and 'there are uncountably many', and more generally all quantifiers of the form 'There are exactly κ many', 'There are at least κ many', and 'There are at most κ many', for κ infinite. In §§5.3–5.6, we shall argue, by bottom-up means, that all these quantifiers are logical. Before that, in §5.2 we tackle a popular argument for first-orderism. As we show, it falls short of establishing that FOL is the one true logic.

---

[3] As discussed in Paseau (2019b) and its sequel Paseau (2021b).
[4] Many authors also impose other conditions on logical quantifiers; van Benthem (1989) is a case in point. These further conditions are usually motivated by linguistic considerations, for example because quantifiers not respecting them are not instantiated in any existing natural language. For a logician striving to capture implicational facts about *possible* extensions of natural language, empirical arguments about what types of quantifiers are or are not instantiated by existing languages are of less pressing concern.

5.2 The concurrence argument for FOL

FOL is mathematically unique in certain ways. No logic with more expressive power than FOL is compact and also satisfies the Löwenheim-Skolem Theorem (Lindström's First Theorem). Equally, no logic with more expressive power than FOL satisfies the Löwenheim-Skolem Theorem and has a recursively enumerable set of validities (Lindström's Second Theorem).[5] These model-theoretic properties, and other similar ones, justify some of the mathematical focus on FOL. And they allow us to put first-order axiomatizations to tremendously useful application in discovering more mathematics. Yet, *pace* Tharp (1975), they do not significantly promote FOL's status as a foundational logic. To say that FOL is the only logic enjoying properties A, B, and C does not advance its cause unless each of A, B, and C is a desideratum on any foundational logic (perhaps because the concept of logical consequence includes the idea that logic must satisfy these constraints, or for some other reason). Not so for the properties just noted. Satisfying the Löwenheim-Skolem Theorems, for instance, is neither an intuitive condition nor one that emerges from established practice. On the contrary, it seems an undesirable feature of a foundational logic, as it limits its ability to discriminate between models of different sizes. Likewise, although there are mathematical grounds for finding compactness attractive—its centrality to first-order model theory is not in question—there are no reasons to insist on it from a foundational perspective. Although a necessary condition for axiomatizability (about which more below), compactness is not in itself desirable.

Several logicians believe that FOL is useful for mathematical purposes whereas SOL is best for foundational ones. Thus John Baldwin in his recent book:

> ...we agree that normal informal mathematical reasoning would most easily be formalized in second-order logic. But in contrast to discussing the foundations of arithmetic and analysis, our focus is on the role of formalization in solving problems of modern mathematics. For us, compactness and categoricity in power [having exactly one model, up to isomorphism, of some cardinality] for first order logic are not deficits but powerful tools for understanding mathematics.
>
> (Baldwin 2018, p. 7; see also p. 34)

Our point of view, as we shall see later, is that neither FOL nor SOL is the one true logic.

Things are otherwise with FOL's axiomatizability, or so it has been thought. A durable argument for FOL as the one true logic, voiced by Quine in his influential *Philosophy of Logic*, is that FOL's special status flows from its possessing a sound

---

[5] Lindström (1969). For textbook proofs, see ch. XIII of Ebbinghaus *et al.* (1994).

and complete proof system, that is, from its axiomatizability. A proof-theoretic conception of logical consequence takes the one true logic to be exhaustively given by a set of inferential rules. A model-theoretic conception, in contrast, defines consequence in terms of truth-preservation over the one true logic's models. Thanks to its completability by a sound proof system, FOL is susceptible to both motivations. In FOL's case, both the proof-theoretic and the model-theoretic conception point in the same direction. This remarkable convergence is a sign, perhaps even proof, of FOL's privileged status. Thus Quine:

> A remarkable concurrence of diverse definitions of logical truth . . . already suggested to us that the logic of quantification as classically bounded [FOL] is a solid and significant unity.   (Quine 1970, p. 91)

However, SOL is not axiomatizable, nor is any infinitary logic worth its salt. The unaxiomatizability of SOL is a standard result, which follows for example from Gödel's First Incompleteness Theorem and the categoricity of second-order Peano Arithmetic.[6] That no (non-trivial) logic closed under infinitary conjunction is axiomatizable is a consequence of the fact that proof procedures are finitary, so any axiomatizable logic cannot but be compact.[7]

In short, many first-orderists are swayed by Quine's concurrence-of-ideas argument. It seems to us, however, that Quine's argument ultimately cannot stand on its own two feet; it presupposes some other criterion of logicality. In brief, the reason is that many logics are axiomatizable; the reason for taking one or another of them to be the one true logic therefore cannot depend on axiomatizability alone.

We illustrate this counter-argument with a specific example. Consider the quantifier 'there exist uncountably many'. By a historical irony, some of this quantifier's key properties were articulated in a long article by H.J. Keisler in 1970, the very same year in which Quine published his book *Philosophy of Logic* in which he advanced the concurrence-of-ideas argument.[8] The quantifier's proof-theoretic properties present a challenge to those first-orderists who, like Quine, see axiomatizability as a key property.

To see this, extend FOL to the logic $\mathcal{L}_Q$ by adding to it the quantifier Q, interpreted as 'there are uncountably many'. More precisely, augment FOL's vocabulary with the symbol Q; supplement FOL's formation rules with the stipulation that if ϕ

---

[6] See §7.1 for the latter.

[7] That is, if a conclusion is derivable from a set of premises then it must be derivable from one of its finite subsets. We take this to be a constraint on proof procedures *sensu stricto*.

[8] Though Quine's argument had been in the air for a long time, and had previously appeared in print too. Kaufman (1985) is a textbook treatment of the original Keisler (1970). Boolos (1975) is an early critique of some of the arguments in Quine (1970). We join company with Boolos (1975, pp. 49–53) and Shapiro (1991, ch. 2) in resisting the idea that logic should be axiomatizable.

is a formula and α is a variable then $Q\alpha\phi$ is a formula; and add to FOL's satisfaction clauses the following stipulation:

$(\mathfrak{M}, g) \vDash Q\alpha\phi$ iff $\{h : (\mathfrak{M}, h) \vDash \phi$ and assignment h agrees with assignment g on all variables other than $\alpha\}$ is uncountable.

When the sentence δ semantically follows from a set of sentences Γ in this extended logic $\mathcal{L}_Q$, we write $\Gamma \vDash_{\mathcal{L}_Q} \delta$. Note that $\mathcal{L}_Q$ has more expressive power than FOL, since the former but not the latter can express the quantifier 'there exist uncountably many'.[9]

We also extend any sound and complete proof system for FOL with the following four rules for the quantifier Q:

$$\frac{Qx\phi}{Qy\phi[y/x]} \quad \text{if } y \text{ is not free in } \phi$$

$$\frac{}{\neg Qx(x = y \vee x = z)} \quad \text{where } y \text{ and } z \text{ are distinct variables from } x$$

$$\frac{\forall x(\phi \rightarrow \psi)}{Qx\phi \rightarrow Qx\psi}$$

$$\frac{\neg Qx\exists y\phi \quad Qy\exists x\phi}{\exists x Qy\phi}$$

The first rule allows us to rename bound variables; it is trivially sound for the $\mathcal{L}_Q$-semantics given earlier. The second rule—which features no premises—is sound for the $\mathcal{L}_Q$-semantics because it states that uncountably many things are not equal to one of two things (or just one thing). The third rule's soundness follows from the fact that if all ϕs are ψs and there are uncountably many ϕs then there are uncountably many ψs (i.e. the quantifier is monotone increasing). Finally, the fourth rule is sound because if a relation has countably many things in its domain and uncountably many in its codomain then some element of its domain must be related to uncountably many elements of its codomain. More succinctly: the countable union of countable sets is countable. If the sentence δ follows from a set of sentences Γ in this deductive system, we write $\Gamma \vdash_{\mathcal{L}_Q} \delta$.

The remarkable result proved by H.J. Keisler (1970) is that the proof system just given is not merely sound, but also complete for countable sets of sentences.[10] More precisely, if Γ is a countable set of $\mathcal{L}_Q$-sentences and δ any $\mathcal{L}_Q$-sentence, then

[9] One implication of the Löwenheim-Skolem Theorem is that if a (single) FOL-sentence has an infinite model then it has a countably infinite model. Hence no FOL-sentence can be true in all and only models with an uncountable domain, unlike the $\mathcal{L}_Q$-sentence $Qx(x = x)$.

[10] Vaught (1964) had given a completeness theorem for $\mathcal{L}_Q$ with a more complicated set of axioms.

$$\Gamma \vDash_{\mathcal{L}_Q} \delta \text{ iff } \Gamma \vdash_{\mathcal{L}_Q} \delta$$

We call FOL$^\omega$ a system of first-order logic with a countably infinite set of variables, individual and predicate constants, and a finite (or countably infinite) and truth-functionally complete set of connectives, with standard syntax and semantics. This allows us to put Keisler's result another way: if the quantifier Q with the above syntax, semantics, and associated rules is added to FOL$^\omega$, the resulting system, which we might call FOL$^\omega(Q)$, is sound and complete (for countable premise sets).

The soundness and completeness of FOL$^\omega(Q)$ may just be the tip of the iceberg. It is an open question whether more than one transfinite cardinality quantifier can be added to FOL$^\omega(Q)$, whilst still preserving soundness and completeness, or even whether FOL$^\omega(Q)$ with the quantifier 'there exist at least $\aleph_2$' is compact.[11] We also observe in passing that the quantifier Q ('there are uncountably many') is definable in SOL, by a standard argument. Q is not definable in $\mathcal{L}_{\omega_1\omega}$,[12] but it is definable in $\mathcal{L}_{\omega_1\omega_1}$ (see §5.3).[13]

But set these extensions of Keisler's result to one side. Its unextended version is enough to dent Quine's concurrence-of-ideas argument for FOL$^\omega$. For if Quine's line of thought is sound and takes us as far as FOL$^\omega$, it should take us at least as far as FOL$^\omega(Q)$.

One might respond to our anti-Quinean argument by pointing out that Keisler's soundness and completeness theorem only holds for *countable* sets of sentences. The deductive system just given (that of FOL$^\omega$ plus the Q rules) is not complete for entailments involving uncountable sets of sentences. If unrestricted $\mathcal{L}_Q$ were complete, it would be compact. Clearly, however, it isn't, as witnessed by entailments such as $\{c_i \neq c_j : i < j < \omega_1\} \vDash Qx(x = x)$ consisting of an uncountable premise set.

But that response sits ill with the concurrence-of-ideas motivation, which takes proof-theoretic considerations as a constraint on the one true logic. Anyone who accepts axiomatizability as a constraint on logic does so because they take the said logic to somehow represent inferential practice. Since that practice ultimately takes place in natural language (including technical vocabulary), it involves only

---

[11] As Väänänen (2004, p. 50) notes, further axiomatizable extensions of FOL using cofinality quantifiers have also been produced.

[12] One way to see this is that if a sentence of $\mathcal{L}_{\omega_1\omega}$ is satisfiable then it must be satisfiable in a countable model. A version of the downward Löwenheim-Skolem Theorem due to Tarski and Vaught is provable for $\mathcal{L}_{\omega_1\omega}$; see Keisler (1971, p. 69).

[13] We further note in passing that FOL with the cardinality quantifier 'there exist infinitely many' cannot be axiomatized (even for countable sets of sentences), a fact first pointed out by Mostowski. An easy way to see this is that soundness and completeness implies compactness and yet the formalization of §5.3's argument $\mathcal{A}$ in FOL plus 'there exist infinitely many' must be non-compact. Another argument for the same conclusion is that if this logic had a sound and complete proof procedure, then any model for the theory made up of the first-order Peano axioms and the claim 'every number has finitely many predecessors' would be isomorphic to the standard model of arithmetic. The resulting theory of arithmetic would then be semantically and deductively complete, contradicting Gödel's First Incompleteness Theorem.

countably many sentences. We, of course, believe applied logic should consider not just (cleaned-up) English but also its possible extensions (see the Prologue), thereby taking in languages with bigger sets of sentences. But friends of axiomatizability regard logic's inferential aspect as constitutive of its role, and so conceive of logic as capturing a relation that an idealized deductive subject can grasp. That is, they conceive of logic as capturing a relation that applies to sentences drawn from a countable set. Our dialectical point thus stands: Quine's concurrence-of-ideas argument takes us beyond FOL. At least one infinitary cardinality quantifier is logical.

Another response to our anti-Quinean argument might be that it uses set-theoretic results in the metatheory to establish $FOL^\omega(Q)$'s axiomatizability. The fourth rule, in particular, assumes a countable instance of the Axiom of Choice: without it one cannot prove that the countable union of countable sets is countable. But should logic depend on set theory in this way? Now, to prove the soundness and completeness of FOL one must use a good deal of mathematics—set theory or something as powerful. It is well-known that the compactness of FOL— provable from its axiomatizability—is equivalent to the Ultrafilter Lemma or the Boolean Prime Ideal Theorem.[14] The metatheory of FOL is therefore steeped in mathematics. So on pain of undermining FOL as well, the response should not be understood as a complaint about how much mathematics is used in the metatheory to establish axiomatizability. It is better understood as saying that the way we persuade ourselves of the fourth Q-rule's validity is inherently set-theoretic.

This response brings out the point we've been trying to illustrate. The reason first-orderists baulk at Q's logicality is not because truths involving it are unaxiomatizable; as explained, Q's properties *are* axiomatizable in a countable extension of first-order logic. What they object to, rather, is Q's apparently set-theoretic content. To be clear, we are happy to take Q as logical, as will emerge over the course of this chapter and book. We may use mathematics to establish some of the Q-rules' validity, but as we will see in Chapter 9, this is unproblematic: the epistemology of inference and the validity of implication are two separate topics. So we disagree with first-orderists' verdict about Q. But the point of the present section is dialectical: it aims to defuse Quine's concurrence-of-ideas defence of FOL^ω. As we have seen, this supposed mark of logicality must ultimately rest on a *deeper* criterion of the logical. First-orderists see FOL^ω as going beyond logic proper, despite its soundness and completeness, because they rely on an independent criterion of logicality. That's the only way they can block the logicality of Q.

---

[14] The first states that every filter on a set can be extended to an ultrafilter, the second that every Boolean algebra has a prime ideal. For proofs of these equivalences, see Chapter 2 of Jech (1973).

The point, in brief, is that there are just too many sound and complete logics to choose from. Some other reason must be found to pick a particular one of them as the true one. Axiomatizability isn't the heart of the matter.

To conclude, we are not tempted by proof-theoretic accounts of logical consequence. We take the weight of evidence as pointing to this relation's freedom from proof-theoretic constraints. As will become clear, the notion of logical truth is for us just as easy to settle in the simple cases as that of quantifier-free arithmetical truth, and at least as difficult in the hard cases as that of ZFC-independent set-theoretic truth. As soon as logic is seen as the science not of inference but of implication, a requirement of axiomatizability falls away. Implicational facts, about which statements imply others, should not be confused with what facts about what we can deduce from what. Inferential facts are merely the tip of the implicational iceberg. True, logical consequence's link to derivability is the reason we got interested in logical consequence in the first place. If we could not recognize whether implications hold in simple cases, we might never have become interested in logic. But this is an epistemic point, and it would be a mistake to think that our epistemic limitations define the bounds of logic.

Still, we recognize that others who broadly accept the model-theoretic perspective feel the atavistic pull of derivability. Quine's concurrence argument for FOL exploits this lingering attraction. Those tempted by Quine's argument must find some way to argue that 'there are uncountably many' is not logical. As we saw, the only way to do that is to rely on some other criterion of logicality. The moral is that an expression should not qualify as logical merely because its rules are part of a sound and complete formal language. Axiomatizability is a side show to the main criterion.

5.3 Beyond FOL: 'there are infinitely many'

We move on now to a more positive argument for infinitary logic. If Γ is any set of sentences of a logic, δ any sentence in its language, and $\vDash$ is the logic's consequence relation, we may characterize the logic's compactness as follows:

If $\Gamma \vDash \delta$ then $\Gamma^{fin} \vDash \delta$ for some finite subset Γ^{fin} of Γ.

A key metalogical difference between logics such as propositional and first-order logic on the one hand and second-order or typical infinitary logics on the other is that the former are compact whereas the latter are not.[15]

Now compactness is defined for formal languages equipped with a consequence relation. English, not being a formal language, is not usually considered to be

---

[15] Assuming, as ever, standard/full semantics for second-order logic.

compact or non-compact. But the notion of compactness is easily extended to English and any other natural language. Taking our cue from the formal definition, say that English consequence is compact just when: for S any set of sentences of English and s any English sentence, if S entails s then some finite subset of S entails s. Given that, the question now is: is English consequence compact or non-compact?

We focus on cardinality quantifiers of the form 'there are κ many', which are part of mathematical English and each of which can be expected to be part of some cleaned-up version of the language (see the Prologue). The interest in English's compactness largely stems from its upshot for our choice of foundational logic: the logic that best underwrites logical consequence in English. In keeping with the 'bottom-up' nature of the arguments, we will be careful in this chapter to presuppose only (relatively) uncontroversial characteristics in our account of English consequence. If we assume from the start that English consequence includes instances requiring second-order or infinitary resources or if we assume that certain quantifiers are logical, then it can be a short step to showing that English is non-compact. In this chapter, we do not wish to invoke any assumptions about logical relations or constants (that will be the business of Chapter 6), but be led to conclusions about them by our discussion.

Consider then the following English argument, which we call $\mathcal{A}$ for later ease of reference:

> There is at least one planet.
>
> There are at least two planets.
>
> $\vdots$
>
> There are at least n planets.
>
> $\vdots$
>
> _____
>
> There are infinitely many planets.

Here 'n' ranges over English numerals: 'one', 'two', 'three',..., 'one million', 'one million and one', and so on.[16] Argument $\mathcal{A}$ seems valid, and evidently no finite subset of its premise set entails its conclusion. It seems to follow, then, that English consequence is non-compact. In particular, as is well-known, $\mathcal{A}$'s conclusion is not formalizable by any sentence of FOL, otherwise $\mathcal{A}$'s first-order formalization would be a valid FOL-argument none of whose subarguments with finite premise set is valid. And if $\mathcal{A}$ is valid, so that natural-language consequence is non-compact, it

---

[16] Or 'one', 'one plus one', and so on. We could equally well use Hindu-Arabic numerals: 1, 2, 3,..., 1000000, 1000001, etc. as they are part of E_c^+.

follows that FOL, which *is* compact, cannot be the one true logic. The moral carries over to any natural language into which the argument is translatable and to any other compact logic.

Other philosophers have argued that A or an argument along A's lines witnesses English's non-compactness; see e.g. Boolos (1975, p. 49) or Oliver and Smiley (2013, p. 238). Champions of second-order logic such as Boolos will of course take A to be valid, since it is formalizable as a valid second-order argument. Champions of plural logic would also regard A as valid, for parallel reasons. The latter category includes Boolos, Oliver and Smiley, Yi[17] and many others. Moreover, the literature does not, so far as we know, contain any overt resistance to the idea that A is valid. What the above writers have in common, however, is that they more or less lay down the validity of natural-language arguments formalizable in second-order logic or plural logic (but not first-order logic). But they do not justify the claim that these arguments are logically valid in a detailed and direct way; the arguments' validity falls out only as part of a much broader and more theoretical argument that second-order or plural logic is logic.[18]

Although we too take A to be valid, we believe that this conclusion must be justified with the utmost care and caution. That's something that hasn't been done before, and our aim is precisely to fill this lacuna. To this end, we carry out some work on behalf of the would-be first-orderist. We set out the most plausible objections to A's validity we can think of; and then we answer them. We devote several pages to knocking down these objections, as this is crucial to the success of the bottom-up approach. Our discussion in this section is framed as an objection to first-orderism, the thesis that first-order logic is the one true logic. This is for two reasons: first-order logic is compact, so can stand proxy for any compact logic; and, as indicated earlier, it is a baseline logic, in the sense that no one seriously doubts that *at least* first-order logic is required to capture the validity of English argumentation.

Resistance to A's validity may take one of four forms. The first questions whether English has infinitely many sentences. The second questions whether A is really *logically* valid or valid only in some weaker sense. The third form of dissent is to maintain that A can only be understood under some finite description. The fourth and final objection is to query whether there is a determinate notion of English consequence. We take these objections in turn, and rebut each of them.

---

[17] See his (2006).

[18] The argument Yi (2006, p. 262) mentions is somewhat different from A but illustrates the same point. His argument has premises interpretable as 'c_n is related to c_{n+1}', where each c_n (for n a natural number) is a name; its conclusion is interpretable roughly as 'There are some things such that for any one of them there is another one of them (which may be equal to the first) such that the first is related to the second'. Yi uses this example to demonstrate the non-compactness of plural logic and evidently regards its natural-language interpretation as logically valid. But he gives no further justification of the natural-language argument's logical validity beyond the fact that it is formalizable as a valid argument in plural logic.

Our conclusion at the end of §5.3, which we generalize later, will be that $\mathcal{A}$ *is* logically valid. This shows that English is not compact; in particular, it sinks first-orderism.

5.3.1 The finitude objection

Does English really consist of infinitely many sentences? It has a finite lexicon (vocabulary), it only allows sentences of finite length, and in practice only sentences shorter than some finite length will ever be uttered, written, or understood. So one might object that $\mathcal{A}$ is not really an argument of English, since its premise set contains sentences longer than any finite bound. In which case, it would follow that English is compact, since the consequence relation of any language with finitely many sentences is trivially compact. As a consequence, the $\mathcal{A}$-based objection to first-orderism runs aground; for there is no such argument upon which to base the objection.

Our response to the finitude objection is threefold. The first is that the finitude objection sits ill with the idea that the one true logic is *at least* FOL. If you think there are only finitely many natural-language sentences, a finite fragment of FOL will suffice to model these sentences' implicational structure. The one true logic will then be this finite fragment of FOL, not FOL itself. And it will not do to suppose that a finite fragment of FOL does the job but that it's simpler to pretend that it's FOL for most intents and purposes. For it is extremely simple to define restrictions of FOL whose sentences are all shorter than some upper bound.

Second, as we explained in the Prologue, our best theory of logical consequence should apply not just to cleaned-up English but to its possible extensions. As applied logicians, our interest should primarily be in the latter. It is scarcely contestable that these extensions contain infinitely many sentences. Incontrovertibly, at least some of them can express argument $\mathcal{A}$.

Finally, observe that linguists, philosophers, and other theorists of language conceive of *actual* natural languages as made up of infinitely many sentences.[19]

---

[19] Perusal of virtually any work on the syntax or semantics of natural language supports this point; a nice example is Keenan and Moss (2016). Here is an early statement by Chomsky: 'We might arbitrarily decree that such processes of sentence formation in English as those we are discussing cannot be carried out more than n times, for some fixed n. This would of course make English a finite state language, as, for example, would a limitation of English sentences to length of less than a million words. Such arbitrary limitations serve no useful purpose, however' (Chomsky 1957, p. 23). In a more recent review article, Hauser, Chomsky and Fitch define a 'Faculty of Language in the narrow sense' (FLN), by which they mean 'the abstract linguistic computational system alone, independent of the other systems with which it interacts and interfaces' (Hauser *et al.* 2002, p. 1571). They have this to say about it: 'All approaches agree that a core property of FLN is recursion, attributed to narrow syntax in the conception just outlined. FLN takes a finite set of elements and yields a potentially infinite array of discrete expressions. This capacity of FLN yields discrete infinity (a property that also characterizes the natural numbers) . . . the potential infiniteness of this system has been explicitly recognized by Galileo, Descartes, and the

These sentences are of finite, but arbitrary, length; hence there are infinitely many of them. More precisely, the set of natural-language sentences is usually specified by a set of recursive procedures, which generate sentences of arbitrary length. For example, all the following are sentences of English:

Your grandparents were tall;
Your great-grandparents were tall;
Your great-great-grandparents were tall;
. . .

Now as a matter of empirical fact, there is some finite number N such that you don't have any great^N-grandparents (which N is the least such may be vague). But that does not affect the point that the infinitely many listed sentences are *bona fide* sentences of English. Mathematics provides numerous further examples, for instance '$1 + 0 = 1$', '$2 + 0 = 2$', '$3 + 0 = 3$', etc. (or if you prefer to use letters instead of numerals: 'one plus zero is one', etc.).

In sum, the finitude objection conflicts with a foundational assumption in linguistics and philosophy of language. It also sits ill with first-orderism and the applied logician's proper task of accounting for logical relations in extended English. For these reasons, we find the finitude objection particularly unpromising.

One may object that infinity should be understood in a potential sense. The argument $\mathcal{A}$, it might be said, contains a potential infinity of sentences and not an absolute infinity. Linguists, it might be added, commit themselves to a potential, and not an absolute, infinity of sentences, as witness the quotation in the previous footnote.

Our reaction to this follow-up 'potentialist' objection is also threefold. First, we doubt that linguistics as a discipline strictly cleaves to a potentialist understanding of the infinite. Mathematical methods are well-established in linguistics, and the standard contemporary conception of the infinite in mathematics is absolutist, not potentialist. Formal semantics is knee-deep in set theory, a theory of absolute infinities *par excellence*. Second, we don't regard the potentialist conception of the

---

17th-century "philosophical grammarians" and their successors, notably von Humboldt... The core property of discrete infinity is intuitively familiar to every language user. Sentences are built up of discrete units: There are 6-word sentences and 7-word sentences, but no 6.5-word sentences. There is no longest sentence (any candidate sentence can be trumped by, for example, embedding it in "Mary thinks that..."), and there is no nonarbitrary upper bound to sentence length. In these respects, language is directly analogous to the natural numbers...' (Hauser *et al.* 2002, p. 1571). We thought this tendency universal until we happened on Ziff (1974) and Langendoen and Postal (1984). The former is a quixotic attempt by a philosopher to challenge the received view (as Ziff calls it) that English has infinitely many sentences. Langendoen and Postal (1984) also challenge the thesis that natural languages have a countable infinity of sentences, but in the opposite direction. They contend that there are English sentences of *transfinite length*, because grammars that place no bounds on sentence length are simpler than those that do. So for Langendoen and Postal, the collection of English sentences is a proper class.

infinite as tenable. Although historically influential, the potentialist movement, if we may call it that, has been on the back foot since the development of set theory in the late nineteenth and early twentieth century. Few philosophers would seriously take the set of natural numbers or of English sentences to be only potentially infinite. To attempt to reject the validity of A on this basis seems at best quixotic. Contemporary philosophers who try to respect potentialist ideas originating from Aristotle, developed in the twentieth century in different ways by Hilbert, on the one hand, and various constructivists, on the other, shy away from denying the actual infinity of cardinalities posited by ZFC. Rather, they apply potentialist ideas to the height of the set universe.[20] Third, the potentialist counterobjection is in any case misguided. For we can simply understand A's premise set in a potentialist spirit and on that basis rerun a version of the same argument. Our point does not hinge on conceiving of A's premise set as absolutely, rather than potentially, infinite.

In sum, the finitude objection fails. Let's turn to a different attempt to block our objection to first-orderism based on A's validity.

5.3.2 The *mathematical* validity objection

For argument A to witness English's non-compactness, it must be a *logically* valid argument. If it is valid only in some weaker sense, it fails to make the antecedent of compactness true,[21] and so fails as a counterexample. There are of course many kinds of non-logical validity, for example metaphysical validity or mathematical validity. Here is an example of a metaphysically valid argument, on Kripkean assumptions: 'The stuff in the glass is water; *so* the stuff in the glass is H_2O'. This argument is truth-preserving of metaphysical necessity but it is not *logically* valid, since the vocabulary on which its validity depends—'water', 'H_2O'—is not plausibly logical. Similarly, the objection goes, the vocabulary on which the validity of A depends—'one', 'two', ..., 'infinitely many'—is not properly logical.

Clearly, A is at least mathematically valid. The word 'planet' in A could be replaced by any other common noun without affecting the argument's validity, so its validity does not turn on any specifically astronomical facts but on the mathematical fact that to be infinite is to be at least as large as $1, 2, \ldots, n, \ldots$. Failure to appreciate A's validity is a mathematical failing, in a way that failure to appreciate the validity of 'The stuff in the glass is water; *so* the stuff in the glass is H_2O' is not a mathematical failing.[22]

---

[20] For an illustration of this tendency, see Linnebo (2013).

[21] That is, S does not logically entail s.

[22] The argument to follow can be extended to other forms of validity by considering the nesting of possible worlds. Much the usual picture is that the logically possible worlds form the broadest class, constrained only by logic. The mathematically possible worlds are a proper subset, constrained not

So might A be merely mathematically valid, and not logically valid? The objection would most likely be that the vocabulary on which its validity depends— 'one', 'two', 'infinitely many'—is not properly logical. We take it that 'infinitely many' is the only plausible source of controversy here. For any finite n, that there are at least n many planets is of course first-order definable (formalizing 'is a planet' as a monadic predicate letter), and we assume that the one true logic is at least as strong as FOL.[23] So, if this objection is to get off the ground, it must be that the quantifier 'there are infinitely many' is not a logical constant.

As noted at the start of this chapter, we must be careful, dialectically speaking. Our argument is 'bottom-up' rather than 'top-down': in the present chapter, we do not want to invoke a general criterion of logical constanthood. To put it another way, we want to argue that A is logically valid on independent grounds and thereby show that 'there are infinitely many' is a logical constant (as we believe it is), rather than presuppose this fact to show that A is logically valid. So if we defend a criterion for logical constanthood that 'there are infinitely many' passes, we risk begging the question in favour of English's non-compactness. This will be the business of the next chapter, but fortunately there is an independent route to that same conclusion.

A is *mathematically* valid, that is, valid assuming the mathematical fact that to be infinite is to be at least as great as 1, at least as great as 2, ..., at least as great as n, ... Our claim is that it is also *logically* valid; and the objection is that it is mathematically but not logically valid. To vindicate our claim, we must exploit a well-known fact about mathematical validity, appreciated by logicians. This is the fact that all mathematically valid arguments may be turned into logically valid arguments from an appropriately specifiable set of mathematical axioms.

All present or recent work in mathematical logic provides evidence for this fact, which is of a broadly inductive nature. What the past hundred-plus years have shown is that there are no irreducibly mathematical steps in argumentation. Mathematical experience teaches that mathematically valid arguments may be turned into logically valid arguments from a set of mathematical axioms pertaining to the domain in question. For example, the truths of arithmetic follow logically from the axioms owed to Dedekind and named for Peano; those of analysis from the axioms characterizing a complete ordered field; those of group theory from the axioms of group theory plus relevant supplementations in a given context (e.g. that a group is abelian or finite); likewise for truths of geometry, topology, set theory,

---

only by logic but also by mathematics. The metaphysically possible worlds are in turn a proper subset of the mathematically possible ones, and thus a further proper subset of the logically possible ones. A is a *mathematically* valid argument, hence truth-preserving in all mathematically possible worlds. Since the mathematically possible worlds are a proper superset of the metaphysically possible worlds, it cannot be objected that A is *merely* metaphysically valid: it is valid in at least as strong a sense as *mathematical* validity. The same argument will extend to any notion of modality whose corresponding possible worlds are a subset of mathematically possible worlds.

[23] Recall that this is understood throughout as first-order logic with identity.

and so on. All branches of mathematics have been axiomatized in such a way that a mathematically valid argument—a mathematical proof—may be cast as a logically valid argument from a set of mathematical axioms that applies to the branch (or branches) in question.

There remains room for disagreement about the logic in which mathematical arguments may all be cast. That is to say, about how to interpret the word 'logical' in the uncontroversial equation 'mathematical validity = logical validity from mathematical axioms'. First-orderists plug first-order logic into this equation: they believe that all mathematical proofs may be converted into first-order proofs from mathematical axioms. The conviction that this can always be done is what various authors, starting with Barwise (following a suggestion of Martin Davis), have called 'Hilbert's Thesis'.[24] Naturally, for the thesis to be substantive, there must be some implicit constraints on which mathematical axioms are to be used. It will not do simply to add the conclusion as a premise every time; the axioms in a given branch must at least be held constant. The axioms that may be used must be in some sense appropriate, although it is difficult to spell out exactly what 'appropriate' means. In practice, however, it is quite easy to recognize it—just look at the countless formalizations logicians have produced of mathematical arguments.

As an illustration of Hilbert's Thesis, first-orderists would contend that elementary set-theoretic arguments, such as $\mathcal{A}$, can all be turned into logically valid ones from the axioms of a first-order set theory (e.g. ZFC). More generally, those who think the correct foundational logic is compact (such as, but not limited to, first-orderists) will take the relevant logic to be compact. In contrast, second-orderists maintain that, at least in some instances, the relevant logic should be second-order logic. To capture all the truths of arithmetic, they argue, we must use second- rather than first-order Peano Arithmetic. To capture those of analysis, we must use a second-order (sub)system of analysis.[25] The mere statement of the axioms of topology requires second-order resources. And so on.

Suppose then that you take $\mathcal{A}$ to be mathematically but not logically valid. It follows from what we have said that $\mathcal{A}$ can be rendered logically valid by the addition of a mathematical premise or premises. Now this additional premise (or premises) must express in some way the idea that to be larger than any finite size is to be infinite. Let's assume that P is such a claim (or a collection of claims to the same effect). The argument $\mathcal{A}$ supplemented with the required extra mathematical premise(s) becomes:

---

[24] See Barwise (1977, p. 41).
[25] See for example the Reverse Mathematics programme outlined in Simpson (2010).

There is at least one planet.

There are at least two planets.

$\vdots$

There are at least n planets.

$\vdots$

P

_____

There are infinitely many planets.

Call this augmented argument $\mathcal{A}^P$.

Now notice: our discussion may be rerun on the supplemented argument $\mathcal{A}^P$. By assumption, $\mathcal{A}^P$ is logically valid. Yet no finite subset of $\mathcal{A}^P$'s augmented premise set implies its conclusion, since the claim (or claims) P together with finitely many claims of the form 'There are at least n planets' for finite n does not imply that there are infinitely many planets. We are still left with an argument witnessing English's non-compactness. This counterexample is not the original argument $\mathcal{A}$ but its supplementation $\mathcal{A}^P$.

To sum up the discussion: the objection we considered was that $\mathcal{A}$ is mathematically but not logically valid. Plainly, $\mathcal{A}$ is mathematically valid. Now, mathematical experience shows that mathematical validity is equatable to logical validity from appropriate mathematical axioms. But any supplementation of $\mathcal{A}$ by appropriate mathematical axioms will turn it into an argument that is logically valid but none of whose finite sub-arguments is valid.[26] So we are still left with a logically valid argument exemplifying English's non-compactness.

5.3.3 The finite-description objection

The third line of resistance is that we cannot understand argument $\mathcal{A}$ except via some finite description. This description might be along the following lines:

The argument $\mathcal{A}$ consists of the premises 'There are n planets' for each finite number n and the conclusion 'There are infinitely many planets'.

What's really going on when we understand $\mathcal{A}$, it is urged, is that we understand its finite description. To appreciate $\mathcal{A}$'s validity, we then reason in a finite way using this finite description. (Perhaps using set theory, arguing from the fact that,

_____

[26] To be clear, by a finite sub-argument of $\mathcal{A}$ we mean an argument with the same conclusion as $\mathcal{A}$'s and whose premise set is a finite subset of $\mathcal{A}$'s premise set.

according to the argument, the set of planets is of size at least as great as any natural number to reach the conclusion that this set is of infinite size.) We thereby convince ourselves of A's validity using finite reasoning.

This third objection is a red herring. We may grant its contention, which is that we understand A via a finite description and convince ourselves of A's validity by finite reasoning. That does not change the fact that A is valid but that no finite subset of its premise set entails its conclusion. Arguments of English may consist of infinitely many premises even if we must employ finite terms to characterize them. A's status as a witness to English's non-compactness is unaffected.

It is worth remembering that a logic, in our book, aims to capture implicational facts (see the Prologue). It is not a theory of what we understand when we grasp an argument's validity (or invalidity), or how we come to appreciate such facts. Whether A's premises entail A is *not*, in other words, an anthropocentric question; it is a question about statements and the relations between them. The finite-description objection thus confuses an epistemological fact with a logical one. It confuses how we convince ourselves of A's validity with whether the logical-implication relation obtains between premises and conclusion.[27]

5.3.4 The indeterminacy objection

The final line of resistance queries whether natural language has a determinate consequence relation. According to this objection, there are only various consequence relations that arise from looking at English through a particular theoretical lens; none is the correct one. The above argument for the non-compactness of English assumes, falsely, that there is a determinate notion: *the* English consequence relation. Logical pluralists all take this line; indeed doing so seems tantamount to pluralism.

Part I raised some problems for logical pluralism. We saw that the position is unpromising for a variety of reasons. As well as the particular problems for Beall and Restall's and Shapiro's logical pluralisms, we raised a general worry. Briefly, logical pluralists should want to do some metalogical reasoning. At this level, what logic should they use? If they endorse virtually *all* logics, like Shapiro, the metalogic

---

[27] A related, non-epistemological, version of this objection is that the argument A is *grounded* in some way in a finite argument, e.g. an argument with a premise stating that for all n, 'There are at least n planets' is true, together with compositional truth principles. On a thumbnail, our response to such an objection is twofold. First, we are not aware of any worked-out and plausible accounts of what it is for some arguments to ground other arguments. Accounts in which some kinds of arguments can be concatenated to derive other kinds do not seem to be instances of grounding as the notion is here intended. Second, if you believe in the idea of some arguments grounding others, you are very likely to think that arguments using semantic notions such as truth or semantic devices such as quotation are grounded in arguments that do not use them. Comparing A to an argument B whose premises include the claim that A's premises are true, A is the better candidate for the argument that does the grounding and B for the argument that is grounded.

will be hopelessly weak. If they endorse just some logics, like Beall and Restall, they are relying on an unmotivated view of logical consequence.

In any case, a more direct response to the objection is also available, which does not presuppose monism. Any account of natural-language consequence from the perspective of a particular logic $\mathcal{L}$ will yield a theoretical account of English consequence. In effect, such an account analyses or replaces the informal notion of consequence with the more precise notion of consequence$_{\mathcal{L}}$, the subscript indicating the way in which consequence is now being understood. For example, a first-orderist such as Quine would take the notion of consequence to be that of consequence$_{\mathsf{FOL}}$—consequence as modelled by FOL. Take, then, the notions of consequence$_{\mathcal{L}}$ and validity$_{\mathcal{L}}$ as informed by logic $\mathcal{L}$. Is $\mathcal{A}$ valid$_{\mathcal{L}}$ or not?[28] If yes, then on this conception English is non-compact, ruling out $\mathcal{L}$'s identification with FOL. If no, an explanation is owed as to why $\mathcal{A}$ appears valid despite its invalidity$_{\mathcal{L}}$. What form could such an explanation take?

In response, one could try to deploy one of the first three objections, in §§5.3.1–5.3.3 above. But as we have seen, none of them succeeds. So we have yet to find a way to explain away the appearance of $\mathcal{A}$'s validity. To simply say that $\mathcal{A}$ is invalid because it can't be formalized in $\mathcal{L}$ may be chalked up as a reason to expand one's foundational logic(s) beyond $\mathcal{L}$. In particular, to say that $\mathcal{A}$'s conclusion cannot be first-order formalized fails to explain why $\mathcal{A}$ appears valid; it merely highlights a shortcoming of first-order logic.

5.3.5 Diagnosis

We have seen no good reason to doubt that the argument $\mathcal{A}$ (or some supplementation of it) witnesses English's non-compactness. If the FOL-undefinable quantifier 'there are infinitely many' is logical then FOL cannot be the one true logic. This is, of course, but a small step into the transfinite as far as cardinality quantifiers go. But it opens the floodgates, as we are about to see.

We were careful to argue for this conclusion without resting on any controversial assumptions about the nature of the logical constants. We assumed only that first-order logic is part of logic and did not appeal to criteria of logical constanthood such as topic-neutrality or generality. Our argument is consonant with such criteria, which support the logicality of the quantifier 'there are infinitely many'. (For the case of topic-neutrality, see Chapter 6.) But it does not presuppose them.

---

[28] We omit consideration of theoretical perspectives informed by $\mathcal{L}$ according to which $\mathcal{A}$ is neither valid$_{\mathcal{L}}$ nor invalid$_{\mathcal{L}}$, since this does not correspond to any live options in the literature. We also assume that the claim that no finite subset of $\mathcal{A}$'s premise set entails its conclusion is unimpeachable, whatever one's logical orientation.

Where has FOL gone wrong? Why can't it capture $\mathcal{A}$'s validity? At the root of FOL's inability to capture logical relations between English sentences is the fact that some of these sentences are involved in infinitary implications. The corresponding arguments, expressible in current natural languages, afford us glimpses of a much more extensive implicational network associated with languages of larger size and longer sentences. In particular, a word such as 'infinite', which appears in $\mathcal{A}$, packs in infinitary content: it is equivalent to the infinitary conjunction of 'at least 1', 'at least 2', and so on.[29] Natural languages allow for infinite conjunction of this kind not by syntactic but semantic means. FOL, however, does not sanction infinite conjunctions, whereas infinitary logics do. In particular, 'there are infinitely many' is $\mathcal{L}_{\omega_1\omega_1}$-definable, by the sentence

$$\exists\{x_i : i \in \omega\}(\bigwedge_{i<j<\omega} x_i \neq x_j),$$

in which an ω-length conjunction of inequations is preceded by an existential quantifier over ω-many variables. Or we could take the infinite conjunction $\bigwedge_{i\in\omega} \exists_i$ of FOL-sentences in $\mathcal{L}_{\omega_1\omega}$, in which $\exists_i$ abbreviates the usual first-order rendering of 'there are at least i things'. And we can, in a similar vein, define 'there are infinitely many Fs' in these logics. From this point of view, $\mathcal{A}$'s logical validity is easily explained: it is simply a countably infinite form of conjunction introduction.

Other English words pack in infinitary conjunctive or disjunctive content, a touch more subtly than the word 'infinite'. For example, the word 'ancestor' is equivalent to the infinitary disjunction 'is a parent', 'is a grandparent', 'is a great-grandparent', and so on. That's why the following argument is conceptually (but not formally) valid:

Al is not my parent.

Al is not my grandparent.

Al is not my great-grandparent.

$\vdots$

Al is not my ancestor.

[29] The phrasal verb 'pack in' is not intended to suggest a philosophy of logical atomism, whereby the word 'infinite' is in some literal sense analysed as a conjunction. We are merely drawing attention to logical connections.

To turn this argument into a logically valid one requires adding the premise that an ancestor is a parent, or a grandparent, or a great-grandparent, etc. In a logic which allows countably infinite disjunction, the validity of the thus-augmented argument is easily captured. Its formalization would be:

$$\neg P_1 a$$
$$\neg P_2 a$$
$$\neg P_3 a$$
$$\vdots$$
$$\forall x (Ax \leftrightarrow \bigvee_{i \in \omega} P_i x)$$

$$\overline{\qquad\qquad\qquad}$$

$$\neg Aa$$

Similar points could be made about 'descendant' (the infinite disjunction of 'child', 'grandchild', etc.), 'is an eventual successor of' (the infinite disjunction of 'succeeds', 'succeeds the successor of', etc.), and several other expressions. Mathematics is replete with examples of this sort: 'polygon' in geometry, 'polynomial' in algebra, 'connected' in graph theory, and numerous others.

In sum, we don't have to go as far as extended English (though we will also do that later) to see that FOL's finitude is responsible for its implicational failings. Natural language as it stands today shows us, through a glass darkly, that the one true logic must be infinitary. Pursuing the Pauline allusion, 'face to face' knowledge of infinitary logics lays bare the true nature of such implications: their conclusions are simply equivalent to the (respective) infinitary conjunction or disjunction of the argument's premises.[30] An infinitary logic transparently vindicates $\mathcal{A}$'s validity.

Naturally, friends of SOL will beg to differ. SOL comfortably formalizes both the quantifier 'there are infinitely' and the notion of an ancestor. The notion of infinity is SOL-definable by a sentence such as

$$\exists R(R \text{ is functional} \land R \text{ is injective} \land \neg R \text{ is surjective})$$

where R is a binary predicate variable. This sentence states that the domain is Dedekind-infinite (i.e. admits an injective but not surjective map onto itself), hence is infinite.[31] The SOL-definition of the quantifier 'there are infinitely many'

---

[30] Paseau (2021a) has more on theism and infinitary logic.
[31] 'R is functional' abbreviates $\forall x \exists! y Rxy$, 'R is injective' abbreviates $\forall x \forall y \forall z ((Rxz \land Ryz) \rightarrow x = y)$ and 'R is surjective' abbreviates $\forall y \exists x Rxy$.

(and so 'there are infinitely many *Fs*', etc.) follows the same pattern. The notion of an ancestor is also easily SOL-definable, as we know from Frege (see also §7.1). But as we are about to see in the rest of this chapter, SOL fares less well when it comes to generalizations of the present argument against FOL.

5.4 Generalization: 'there are κ many'

We saw in §5.3 that argument $\mathcal{A}$ is valid, so is a witness to English's non-compactness. Hence the one true logic cannot be FOL, nor any other compact logic. How far does this point generalize?

In the Prologue, we explained why our account of logical consequence should apply not just to cleaned-up English but to its extensions. English, being a countable language, can only define countably many quantifiers of the form 'there are κ many'.[32] It can define all finite quantifiers of the form 'there are n many' for finite n, and countably many of the form 'there are κ many' for κ infinite. It is hard to see, however, why extensions of English couldn't contain any of the cardinality quantifiers English cannot define. Let κ be a cardinal not definable in English. An extension of English containing 'there are at least κ' is perfectly conceivable, as is the extension containing all quantifiers of the form 'there are at least λ many' for all λ up to and including κ. To cut off possible extensions of English at some particular ordinal and declare that beyond this point there can be no others would be arbitrary.

It would be hard to disagree that for any κ, English augmented with a name for κ (and a corresponding cardinality quantifier) is a possible extension of English. And just as logicians conceive of languages with κ-many constants for large infinite κ, we find it plainly conceivable to extend English with names for all cardinals up to and including κ. The claim that English augmented with a name for *every* κ is a possible extension of English is a much stronger assumption than anything needed here.

So for any uncountable limit cardinal κ,[33] consider a possible extension of English in which all numbers up to and including κ itself are definable.[34] Call $\mathcal{A}^\kappa$ the analogue of the argument $\mathcal{A}$ (in this notation $\mathcal{A}$ would be $\mathcal{A}^\omega$). $\mathcal{A}^\kappa$'s premises are 'There at least λ planets' for all $\lambda < \kappa$ and its conclusion is 'There at least κ planets':

---

[32] Keeping context fixed (see the Prologue). The capacity of English to express cardinality quantifiers can vary with context; consider the indexical 'at least as many as those', said alluding to some things.

[33] That is, a κ such that $\kappa = \aleph_\alpha$ for α a limit ordinal (not equal to 0).

[34] This fragment of extended English is recursively specifiable *given* the atomic clauses, of which there are κ. What such an extension of English would look like seems to us plainly intelligible, even for large κ, given our understanding of the grammar of current English and how names for cardinals behave.

There is at least one planet.

There are at least two planets.

$$\vdots$$

$$\vdots$$

There are at least $\aleph_0$ planets.

There are at least $\aleph_1$ planets.

$$\vdots$$

———————————

There are at least κ planets.

A similar dialectic to that in §5.3 can now be played out for the argument $\mathcal{A}^\kappa$, and most of the objections to it parried in the same way. Possible extensions of English can have κ-many sentences, which dispatches the analogue of the finitude objection. The finite-description objection is similarly a red herring. And the indeterminacy objection must ultimately depend on another objection, as before. The only objection that cannot be batted away in the same way is the mathematical validity objection. This is the objection that sees the conclusion of $\mathcal{A}^\kappa$ as following from its premises by mathematical, rather than logical, necessity. So here we have to make do with a weaker claim, that the infinitary logician's account of the validity of this implication is the best possible. As the infinitary logician sees it, $\mathcal{A}^\kappa$ is valid simply because its conclusion is equivalent to the κ-ary conjunction of its premises. This is the same diagnosis as in the case $\kappa = \omega$, i.e. for the argument $\mathcal{A}$.

The arguments $\mathcal{A}^\kappa$ with κ singular are a special case. In this case, the infinitary form of conjunction need not even be κ-ary. If, for example, $\kappa = \aleph_\omega$ then the conclusion 'There are at least $\aleph_\omega$ planets' is equivalent to the countably infinite conjunction of the premises, the premise set being {'There are at least n planets': n is finite} $\cup$ {'There are at least $\aleph_n$ planets': n is finite}. Observe in passing that the argument $\mathcal{A}^{\aleph_\omega}$ is expressible using current English (including mathematical English); to express its premises and conclusion, we can use the resources of current English, and need not pass to an extension. In the specific case of $\mathcal{A}^{\aleph_\omega}$, then, the reply to the mathematical validity objection proceeds in an exactly parallel way to the reply in §5.3.2.

In sum, the one true logic ought to model possible extensions of current natural language. So it must underwrite the validity of $\mathcal{A}^\kappa$ for all infinite limit cardinals κ. Given the lingering mathematical validity objection, we have not quite proved that last claim in this section; but we hope to have made it plausible. The logic $\mathcal{L}_{\infty\infty}$ underwrites these arguments' validity, although no logic such as FOL or SOL or any sublogic of $\mathrm{FTT}_{\infty\infty}$ of sort $\kappa\lambda$ (or sort $\kappa\infty$ or $\infty\lambda$) can, since any such logic can only define a bounded number of cardinals (see the appendix to this chapter). In particular, none of the standard finitary logics (such as FOL or SOL) can be the one true logic. A stronger corollary is that if the one true logic is a sublogic of $FTT_{\infty\infty}$, its sort must be $\infty\infty$.

5.5 Beyond SOL: infinitary conjunction

Fans of second-order logic are impressed by its ability to handle inferences involving properties and plural expressions, by the continuity of its semantics with that of first-order logic, by its ability to characterize mathematical structures, and by its ability to express various generalized quantifiers (see §7.1 for a review of some of these virtues). We mentioned in §5.4 that there are arguments $\mathcal{A}^{\kappa}$ whose validity SOL does not capture. In this section and the next, we consider two further arguments for going beyond second-order logic (SOL). These arguments are addressed to the growing band of philosophers and logicians who take SOL to be the one true logic. (Our second further argument may be considered a variant of the one in §5.4, as it is closely related to it.) Our main moral will be that the one true logic cannot have finitary sort (sort $\omega\omega$), as SOL does; a stronger moral is that the logic's sort must be $\infty\infty$.

5.5.1 Superhumans

Let's recap Part II's aims. As explained in the Prologue, our objective is to characterize the logical-consequence relation not merely of cleaned-up English (or any other natural language) but of its extensions (E_c^+). Bearing that in mind, let's imagine some beings whose powers transcend ours. Although the thought experiment to follow is based on these beings' greatly enhanced inferential powers, for us it will illustrate an implicational moral.

Imagine if you will *superhumans*, who can accomplish infinite tasks in a finite amount of time. In other words, superhumans are able to perform supertasks. In particular, they are capable of uttering infinitely long sentences such as

0 is even and 2 is even and 4 is even and ... $2n$ is even and ...,

in a finite amount of time. Let's imagine that they do so by speed-up: it takes them a second to say '0 is even', half a second to then add 'and 2 is even', a quarter of a second to further add 'and 4 is even', and so on; so it takes them two seconds to utter the full infinitary sentence. Whether it is a mere medical impossibility for *us* to perform supertasks, as Russell put it,[35] remains unclear. But what *is* fairly

---

[35] Russell (1936, p. 143) considers it medically, rather than logically, impossible to run through the decimal expansion of π.

clear is that such beings as the superhumans are a conceptual possibility. Indeed, some physicists and philosophers of physics have argued that supertasks are even a physical possibility, as their existence is consistent with general relativity.[36]

Suppose then that superhumans perform the following inference

0 is even and 2 is even and 4 is even and ... $2n$ is even and ...

---

0 is even and 4 is even and 8 is even and ... $4n$ is even and ...

The inference is an infinitary form of conjunction elimination: it drops the premise's every other conjunct, starting with the second. (The dropped conjuncts are '2 is even', '6 is even', and so on for all even numbers that are not multiples of four.) The inference cannot be modelled in propositional, first-order, or second-order logic, nor in any logic not closed under countable conjunction. To be clear, the premises and conclusion contain names and no quantifiers; in particular the conclusion cannot be formalized as for example 'for all n, if n is a multiple of four then n is even'. The superhumans' inference seems patently logical.

To reinforce this verdict, we may imagine superhumans' practice in more detail. They use the connective 'and' as we do and perform the finitary inferences we do in all finitary contexts. Their inferences in the finitary case conform to the usual finitary introduction and elimination rules for conjunction, performable by human logicians. If prompted, superhumans will happily explain that the truth-value of a conjunction, finitary or infinitary, is true if and only if all its conjuncts are true. Recognizing that we mere humans are unable to utter infinitely long sentences, and that our auditory powers fail to keep up with them as they speed through an infinite utterance in two seconds, they might patiently explain to us in finitary terms why the above inference is valid. That explanation is the same as the one we provided above, writing as two humans for a presumed human readership. Superhumans warn us, though, not to confuse the argument's finitary description with the inference itself. Doing so, they caution, would be like confusing the inference $\mathfrak{I}$ in which we utter the entire premise '0 is even and 2 is even and 4 is even and ... 1,000 is even' and its conclusion '0 is even and 4 is even and 8 is even and ... 1,000 is even' with a succinct description of $\mathfrak{I}$. The inference $\mathfrak{I}$ might

---

[36] See Earman (1995, ch.4). Consider the task of checking all instances of the Goldbach Conjecture (GC), which states that every even number greater than 2 is the sum of two primes. Agent A and her computer are initially at a certain starting point p_S. A sets off along a worldline with a finite elapsed time, ending up at the endpoint p_E. A's computer sets off along a worldline with an infinite elapsed time, and has been programmed to check all the even numbers for the Goldbach property (being the sum of two primes), one per minute say. If the computer finds a counterexample to GC, it sends a signal that will reach A at p_E. Via her computer, A has thus performed a supertask: GC is true iff A has received no signal at p_E. Whether these kinds of 'bifurcated supertasks', even if they are compatible with General Relativity, can truly be realized in universes like ours is apparently an open question.

take a normal human being (*not* a superhuman) several minutes to utter, whereas its abbreviated description would in contrast take a mere few seconds. Fortunately for superhumans, their infinitary inferences don't take *them* very long. Yet that does not prevent them from appreciating the difference between *performing* an infinitary inference and *describing* it in finite terms. The performance/description distinction is the use/mention distinction, or something very much like it, familiar from the philosophy of language.

It seems as plausible as anything in this area that the superhumans' inference is logically valid. A complete account of logical consequence must therefore underwrite the inference as logical. It should more generally underwrite any inference that deploys a countably infinite form of conjunction elimination: the inference from the premise $\bigwedge_{i \in \omega} P_i$ to $\bigwedge_{i \in S} P_i$ is logically valid whenever S is a non-empty subset of ω. It follows that the one true logic must be infinitary: it must be capable of underwriting countably infinite conjunction elimination.

We may also imagine superhumans performing a supertask to *introduce* a constant. They take a second to utter the first premise P_1, half a second to utter the second premise P_2, a quarter of a second to utter the third premise P_3, and so on. They then utter the inference marker 'so' before embarking on a one-second utterance of the conclusion. To do so, they take half a second to utter P_1 followed by the word 'and', then a quarter of a second to utter P_2 followed by the word 'and', and so on. Superhumans are thus capable of performing countably infinite conjunction introduction. Once more, we can imagine them describing this inference to us in finite terms—as we just have in this paragraph—and warning us not to confuse this short finitary description with the performance of the supertask. The only difference between superhumans and us is that they are endowed with the ability to speed up whereas we are not. This is what allows them to utter an infinite conjunction of statements whereas we can only utter a finite one. In other respects, their practice is identical to ours, and it is hard to resist the conclusion that their form of conjunction introduction is just as logical.

SOL, however, is in general incapable of handling countably infinite conjunction elimination and countably infinite conjunction introduction (as we prove in the appendix to this chapter). This is true more generally of any sublogic of FTT (finite type theory), since all these logics' sort is $\omega\omega$. The one true logic that encompasses all conceivable, as well as actually manageable, inferences must therefore go beyond SOL or any sublogic of FTT. It must be closed under countable conjunction introduction and countable conjunction elimination.

You will have noticed that the facts about superhumans just appealed to are inferential, whereas we are interested in implicational facts (recall the Prologue). Inferential facts only matter in so far as they mirror implicational ones. Our thought experiment is cast in terms of superhumans for vividness. It is a pedagogical device to help us see the validity of the argument and nothing more. It helps us mere humans recognize the infinitary arguments superhumans deploy as valid.

5.5.2 Generalization

How far to generalize the point just made remains unclear. Can we imagine *supersuperhumans*, who can utter a longer-than-$\aleph_0$ conjunction, and perform infinitary conjunction elimination on it? For example, in a world in which time is modelled by the continuum, supersuperhumans might utter a conjunct from a $2^{\aleph_0}$-sized premise set at each instant in the time interval $[0,1)$ say, pause for a (literal) instant after the second is up, and then utter a conjunct succeeded by 'and' from the $2^{\aleph_0}$-length conclusion at each instant in the time interval $[1,2)$.[37] We may suppose that the conclusion contains a subset of the premise's set of conjuncts, so that the resulting argument is valid. An example: the premise set is the set of all sentences of E_c^+ of the form 'x is a number' for x a real number in $[0,1]$, and the conclusion is the conjunction of all sentences of the form 'x is a number' where x is in $[0,1]$ and its decimal expansion contains infinitely many 5s. The rule of inference supersuperhumans employ is thus continuum-sized conjunction elimination.

It is hard to see why supersuperhumans couldn't exist (in the broadest sense of objective possibility). *We*—the authors of this book—find their existence perfectly imaginable. Although we are both well aware of the fact that imaginability does not entail possibility, nothing seems to stand in the way of the entailment here. Supersuperhumans' existence certainly does not contravene any metaphysical tenet. Naturally, just like superhumans, supersuperhumans are a helpful heuristic. The infinitary arguments they deploy are valid because they are valid in the obvious infinitary semantics, not because supersuperhumans perform them. The thought experiment based on their extraordinary inferential prowess is meant to help us, including readers agnostic on the point, discern some infinitary implicational relations.

As far as we can tell, then, continuum-sized conjunction elimination is a logically valid inference. The same story can be told for the logicality of continuum-sized conjunction introduction, disjunction introduction and disjunction elimination. Now our best set theory of cardinality puts no upper bound on the value of $2^{\aleph_0}$: the range of α for which $2^{\aleph_0} = \aleph_\alpha$ is consistent with ZFC (or any widely accepted extension) is unbounded.[38] So it is epistemically open that possible subjects perform κ-ary conjunction elimination, for any κ from an unbounded class of cardinals.[39]

---

[37] The interval $[0,1)$ consists of all real numbers from 0 to 1, including 0 but excluding 1; similarly for $[1,2)$.

[38] König's Theorem merely informs us that $2^{\aleph_0}$'s cofinality must be uncountable.

[39] The scope of the operators matters here. To avoid misunderstanding, our claim is of the form 'for any κ from an unbounded class, it is epistemically open that...' rather than 'it is epistemically open that for any κ from an unbounded class, ...'.

A possible objection to the supersuperhumans thought experiment is that theirs isn't a proper *language*. In particular, if their 'language' is uncountable, then there won't be a recursive specification of its syntax in a way that languages require. But why demand that languages be capable of recursive characterization? The usual reason, probably most associated with Davidson (1965) is that, otherwise, the language isn't *learnable*. This is 'how an infinite aptitude can be encompassed by finite accomplishments' (1965, p. 8). When we are dealing with finite creatures, such as humans, this seems a sensible demand. But we are here dealing with *supersuperhumans*, who are not so limited. There seems no good reason to believe that languages must be recursively specifiable in order for a supersuperhuman to be able to learn them.

The conclusion is that the one true logic's sort cannot be finitary. The logic must be closed under countable conjunction ($\aleph_0$-conjunction), as the superhumans' thought experiment shows. Not only that, it must also be closed under at least $\aleph_1$-conjunction, as shown by the supersuperhumans' thought experiment, and perhaps under a much larger infinitary-sized conjunction, depending on the value of the continuum. Stripping the arguments of their inferential gloss, the more general point is that the one true logic is closed under such conjunctions. And it is hard to see why, if $\aleph_1$-sized conjunctions of propositions are acceptable, $\aleph_2$-sized, or $\aleph_3$-sized, etc. conjunctions should not be.

We are aware that the arguments in this section are not watertight, as is usually the way with thought experiments. But we take them to be compelling and to strengthen the conclusion that the one true logic's sort is maximally infinitary.

5.6 Beyond SOL: cardinality quantifiers

We turn to a final argument for transcending second-order logic, closely related to the one in §5.5 based on the argument whose conclusion was 'there are at least κ many things'. As mentioned, SOL defines only a countable infinity of cardinality quantifiers (see the appendix), so there must be a least undefinable such. Friends of SOL take all these cardinality quantifiers to be logical. The flip side for the second-orderist (anyone who takes SOL to be the one true logic) should be clear. She must deny that all but the countable infinity of SOL-definable cardinality quantifiers are logical. This is the great majority of cardinals—indeed all of them but a countable infinity.

Although we see no decisive argument against the second-orderist's position, it strikes us as unattractive, bizarre even. The SOL-definable quantifiers are scattered throughout the ordinal spine in a way that does not seem to track logicality at all. All the finite cardinality quantifiers and a countable initial segment of the transfinite ones are logical. Beyond that, some of the quantifiers are logical, including very large ones such as 'there are I-many' or 'there are M-many' where

I and M are the least inaccessible and least Mahlo cardinal respectively. The great majority, however, are not logical, according to the second-orderist. Let α be the least ordinal such that 'there are at least $\aleph_\alpha$' is not SOL-definable, and let β be the next limit ordinal after α such that $\aleph_\beta$ is SOL-indefinable. As the second-orderist sees it, a claim such as 'If there are at least $\aleph_{\omega_1+1}$ things then there are at least $\aleph_{\omega_1}$ things' is a logical truth (since it is formalizable as an SOL-validity), whereas 'If there are at least $\aleph_{\alpha+1}$ things then there are at least $\aleph_\alpha$ things' is *not* a logical truth, since it is not SOL-formalizable. Here is a table illustrating the strange shape of the second-orderist's commitments:

Statement	Logical Truth?
If there are at least 2 things then there is at least 1 thing	Yes
If there are at least 3 things then there are at least 2 things	Yes
. . .	. . .
If there are at least $n+1$ things then there are at least n things	Yes
. . .	. . .
If there are at least $\aleph_1$ things then there are at least $\aleph_0$ things	Yes
If there are at least $\aleph_2$ things then there are at least $\aleph_1$ things	Yes
. . .	. . .
If there are at least $\aleph_{\omega+1}$ things then there are at least $\aleph_\omega$ things	Yes
. . .	. . .
If there are at least $\aleph_{\alpha+1}$ things then there are at least $\aleph_\alpha$ things	No
. . .	. . .
If there are at least $\aleph_{\beta+1}$ things then there are at least $\aleph_\beta$ things	No
. . .	. . .
If there are at least $\aleph_{\omega_1+1}$ things then there are at least $\aleph_{\omega_1}$ things	Yes
. . .	. . .
If there are at least $\aleph_{I+1}$ things then there are at least $\aleph_I$ things	Yes
. . .	. . .
If there are at least $\aleph_{M+1}$ things then there are at least $\aleph_M$ things	Yes
. . .	. . .

Thinking about the superhumans introduced in §5.5 lets us bolster the present argument. The ordinal α is countable and thus has cofinality ω: there is an increasing ω-sequence of ordinals $\alpha_0, \alpha_1, \ldots \alpha_n \ldots$ whose limit is α. Since all these ordinals are smaller than α, the quantifier 'there are at least $\aleph_{\alpha_n}$-many' is SOL-definable for each (finite) n. By uttering the following ω-sequence of sentences in finite time

$$\text{There are at least } \aleph_{\alpha_0}\text{-many planets.}$$

$$\text{There are at least } \aleph_{\alpha_1}\text{-many planets.}$$

$$\vdots$$

$$\text{There are at least } \aleph_{\alpha_n}\text{-many planets.}$$

$$\vdots$$

a superhuman could thereby make a series of claims collectively equivalent to 'There are at least $\aleph_\alpha$-many planets'. All the premises are SOL-definable but the conclusion is not. If superhumans are possible, their apparently kosher logical practice calls for formal resources going beyond SOL.

We recognize that the argument in this section cannot compel the second-orderist to give up her view. She may stick to her guns and say that a cardinality quantifier is logical just when it is SOL-expressible. End of story, she might insist. We cannot accuse her of inconsistency. The resulting position is not inconsistent; it is merely implausible.

The second-orderist might try another tack. She might dig in her heels and respond that it is only an illusion that we can define α for example. But that is particularly implausible, as its definition is simply: $\aleph_\alpha$ is the smallest cardinal not definable in second-order logic (meaning that the associated cardinality quantifier is not definable). Now in the wake of definability paradoxes such as 'the smallest number not definable in fewer than 100 words', we know that not everything that purports to be a genuine definition really is one. Although the details are disputed, the root cause of such paradoxes is clear enough: the notion of definability in a language is being deployed within the language itself to create trouble. Yet nothing of the sort seems to be going on when we define, in English, the least SOL-indefinable cardinal $\aleph_\alpha$ (this α being a countable ordinal). SOL is a well-behaved and well-understood logic. We can refer to the least indefinable cardinal in this logic, just as we can define the least FOL-indefinable cardinal (viz. $\aleph_0$).

One could respond that this sort of argument will overgeneralize. Many logics are like SOL in that they allow us to define some cardinality quantifiers but not all cardinals smaller than some definable ones. (First-order logic, note, is an exception, as it defines all and only the finite cardinality quantifiers. So its definable cardinality quantifiers are an initial segment of the cardinal hierarchy.) Couldn't the same argument be run against any such logic? Wouldn't that show that no such logic can be the one true logic? The response is that it could and that it does! The correct reaction to the argument's generalizability should *not* be to look at it askance, as if it somehow proves too much. Instead, we should embrace its conclusion and recognize *all* cardinality quantifiers as logical. This paves the way, of course, for $\mathcal{L}_{\infty\infty}$ to be the one true logic—if the logic's type is that of FOL. More generally, the one true logic has sort $\infty\infty$ if it is a sublogic of $FTT_{\infty\infty}$. In any such logic, all cardinality quantifiers, finite and transfinite alike, are definable.

5.7 Conclusion

This chapter has offered some 'bottom-up' arguments for our infinitary hypothesis. We saw that neither FOL nor SOL can account for logical phenomena involving statements of natural language as well as its extensions. We parried an argument for FOL, and argued that the one true logic should allow for quantification over

κ variables and for κ-ary conjunction and disjunction, for any infinite κ. These arguments eschewed highly theoretical assumptions about the nature of logic. In Chapter 6, we consider the theoretical—'top-down'—case, which does proceed from the nature of logic. We then draw some further conclusions in Chapter 7.

Appendix: technical results

We state more precisely the main technical results appealed to in this chapter and prove them. These proofs will be argument sketches rather than full-dress demonstrations. The formal theory of SOL-satisfaction which we rely on and which can be stated in a metatheory of sets will remain implicit. Roughly, we have in mind a background set theory such as second-order ZFC augmented by the assumption of unboundedly many inaccessible cardinals plus resources to define the satisfaction relation between models and formulas of the language.[40] It is also worth pointing out at the outset that our attitude to the universe of sets is a realist one. It follows immediately from this, for example, that terms of a countable language fail to denote all but countably many of the sets in this universe.[41] We come back to this point in Chapter 10.

We start with SOL before moving on to infinitary logics. We mentioned earlier the familiar fact that FOL can only define 'there are n many Fs' whenever n is finite and similar finite cardinality quantifiers (e.g. 'there are at least n many Fs') and others that may be defined in terms of them (e.g. 'there are at least m many Fs or there are at most n many Fs' with $m \neq n$). It *cannot* define 'there are infinitely many Fs' or 'there are at least κ many Fs' or 'there are at most κ many Fs' or 'there are exactly κ many Fs' for infinite κ. In other words, no FOL-sentence is true in all and only models in which infinitely many objects in the domain have the property denoted by F, and likewise for the other expressions. For brevity, when we say that a logic can (or cannot) define a particular cardinality quantifier 'there are κ many' (where κ is a cardinal), what we mean is that it can (or cannot) define the sentence 'there are κ many Fs' by a single sentence of its language. This in turn means that there is (or is not) a single sentence ϕ in the logic true in all and only models in which 'there are κ many Fs' is true. Similarly for 'there are least κ many' and the like.

In the sense just specified, SOL can only define a countable infinity of cardinality quantifiers. This is the result we appealed to earlier in this chapter, which we shall now prove. As we saw in §5.3, SOL can express the quantifier 'there exist a countable infinity of Xs', by a formula which we may call $\phi_{\aleph_0}(X)$. It can also express the notions of being of smaller size than, or being of no greater size than, or being of the same size as. We may respectively abbreviate these as $X < Y$, $X \leq Y$, and $X \sim Y$. It follows that SOL can express 'there exist $\aleph_n$-many' for all finite numbers n by exploiting the following recursive definition:

$$\phi_{\aleph_{n+1}}(X) = \exists Y(Y < X \wedge \phi_{\aleph_n}(Y) \wedge \neg\exists Z(Y < Z < X))$$

A similar argument shows that if $\aleph_\alpha$ is SOL-definable then so is $\aleph_{\alpha+1}$. Observe that SOL can also express 'there exist $\aleph_\omega$-many' in part because it can express the notion of being of size $\aleph_\lambda$ for λ some limit ordinal. The definition of 'X's size is a limit cardinal' is simply

---

[40] A fuller account of some key points about SOL made below may be found in Väänänen (2012). Shapiro (1991, pp. 134–7) has the brief details on one approach to SOL's metatheory.

[41] As opposed to, say, in a countable transitive model of set theory, in which every element is definable by a formula.

$Lim(X) = \forall Y(Y < X \rightarrow \exists Z(Y < Z < X))$

Using $Lim(X)$ and $\phi_{\aleph_0}(X)$, one may straightforwardly define the notion $\phi_{\aleph_\omega}(X)$.

Now letting the superscript indicate the number of symbols in the language, let's call second-order logic with a countably infinite vocabulary SOL^ω. As well as a finite number of truth-functional connective symbols (that are collectively expressively adequate), the existential quantifier and parentheses, SOL^ω contains a countable infinity of each of the following: first-order variables, constants, predicate constants of each adicity, function symbols of each adicity, predicate variables of each adicity, and functional variables of each adicity. Thus the set of sentences (or formulas) of SOL^ω is countable, from which it follows that there is a least ordinal α such that $\phi_{\aleph_\alpha}(X)$ is not expressible in SOL^ω. (This argument could be formulated in a sufficiently strong metatheory.) Notice that $\omega < \alpha < \omega_1$, where ω_1 is the first uncountable ordinal. The reason $\alpha > \omega$ is that SOL^ω can define 'there are infinitely many' and can also evidently define all the first-order finite cardinality quantifiers 'there are n' for finite n. The reason $\alpha < \omega_1$ is that SOL^ω cannot define all the quantifiers 'there are $\aleph_\alpha$-many' for $\alpha < \omega_1$, as there are uncountably many of these. We also note in passing that α is a limit ordinal (by the SOL-definability of κ^+ given the SOL-definability of κ), and must have cofinality ω since it is countable.

What about SOL more generally? Let SOL^κ be second-order logic with vocabulary of size κ, for infinite κ. We may suppose that as well as a finite number of truth-functional connective symbols (that are collectively expressively adequate), the existential quantifier and parentheses, SOL^κ contains κ-many of each of the following: first-order variables, constants, predicate constants of each adicity, function symbols of each adicity, predicate variables of each adicity, and functional variables of each adicity. The key point is that any sentence ϕ_λ of SOL^κ that is true in all and only models of size $\geq \lambda$ must be equivalent to a sentence ψ_λ of SOL^ω (in an expanded second-order language encompassing the vocabulary of both SOL^ω and SOL^κ). Thus SOL^κ can define no more cardinality quantifiers than SOL^ω can.

The argument for this claim is perhaps obvious enough: in a nutshell, the reason is that any predicate or individual constants in ϕ_λ might as well be variables, so the result follows. In a little more detail, let's sketch the argument for the predicate constant case (the constant case is analogous). Suppose $\phi_\lambda(F)$ is a sentence of SOL^κ in which the predicate constant F appears. Let $\exists X\phi_\lambda(X)$ be the sentence resulting from replacing all occurrences of F in $\phi_\lambda(F)$ with a monadic predicate variable X distinct from any already appearing in $\phi_\lambda(F)$ and prefacing the resulting formula with $\exists X$. We show that $\phi_\lambda(F)$ and $\exists X\phi_\lambda(X)$ are equivalent. Clearly, if the model $\mathcal{M}$ satisfies $\phi_\lambda(F)$ it also satisfies $\exists X\phi_\lambda(X)$.

Conversely, suppose $\mathcal{M}$ satisfies $\exists X\phi_\lambda(X)$ with S a subset of $dom(\mathcal{M})$ that satisfies $\phi(X)$ in $\mathcal{M}$. It follows that the model $\mathcal{M}^*$, which is just like $\mathcal{M}$ except that the extension of F in $\mathcal{M}$ is S, satisfies $\phi_\lambda(F)$. Hence, by the definition of $\phi_\lambda(F)$, the model $\mathcal{M}^*$ has cardinality $\geq \lambda$. It follows that $(\mathcal{M}, S)$ also satisfies $\phi_\lambda(F)$ because $\mathcal{M}$'s domain is of the same cardinality as $\mathcal{M}^*$'s domain, that is, $\geq \lambda$. This argument, with the i's dotted, proves the equivalence of $\phi_\lambda(F)$ and $\exists X\phi_\lambda(X)$. By repeated application of this argument (and an analogous argument for constants and predicate letters of other adicities), ϕ may be taken to be a formula equivalent to one in which no predicate or individual constants appear. By relabelling the predicate or first-order variables in this last formula, we obtain a SOL^ω-formula equivalent to the SOL^κ-formula ϕ_λ.

In sum, if SOL^κ can define 'there are at least λ-many', then so can SOL^ω. The upshot is that, dropping the now-superfluous superscript, SOL defines a countable infinity of transfinite cardinality quantifiers, since this is obviously true of SOL^ω. By a similar argument, if $\phi^*(X)$ is a formula of SOL^κ with a single free variable—the monadic predicate variable X—satisfied by all and only subsets of a particular size, then $\psi^*(Y)$ is a formula of SOL^ω with a single free variable—the monadic predicate variable Y—that is satisfied by all and only subsets of that same size.

As explained, the first SOL-undefinable cardinality quantifier is 'there are $\aleph_\alpha$- many' for α a limit ordinal between ω and ω_1. Using slightly more sophisticated resources, we can check that second-order logic can define the first inaccessible cardinal. Since Zermelo (1930), it has been known that the smallest model of ZFC^2, second-order ZFC set theory, is $\langle V_I, \in \rangle$, where I is the first inaccessible cardinal.[42] As ZFC^2 is finitely axiomatizable, this allows us to define the notion of being of least inaccessible size:

$$\phi_I(X) = \exists Y \exists R[ZFC^2(Y,R) \wedge \forall Y^* \forall R^*(ZFC^2(Y^*,R^*) \rightarrow Y \leq Y^*) \wedge X \sim Y]$$

Informally, this states that X is of the same size as the domain of the least model of ZFC^2. It's also relatively straightforward to show that SOL defines the notion of a measurable or Mahlo cardinal (cardinals larger than the first inaccessible) and that if it defines κ then it defines $\aleph_\kappa$.[43]

Finally, we note that SOL is not closed under countably infinite conjunctions or disjunctions, a fact we made use of in §§5.5–5.6. To see why this fact is true, suppose for reductio that ϕ were an SOL-sentence equivalent to the set of sentences $\{\exists x F_i x : i \in \omega\}$. (That is, the sentence and the set of sentences have the same set of models.) Note that ϕ is a finite formula and is not a contradiction. Let j be the smallest natural number such that F_j is not in ϕ. Then, by a standard interpolation theorem for SOL, $\phi \nvDash \exists x F_j x$, a contradiction. Or, for another argument, simply notice that SOL can define the countably many quantifiers 'there exist at least $\aleph_\beta$' for ordinal β smaller than the α defined earlier. But by α's definition, SOL cannot define 'there exist at least $\aleph_\alpha$', which is equivalent to the countable conjunction of 'there exist at least $\aleph_\beta$' for $\beta < \alpha$. Similarly, $SOL_{\kappa\kappa}$ is not closed under κ-ary conjunction or disjunction for regular cardinals κ. (Recall that $SOL_{\kappa\kappa}$ is the logic of SOL's type and of sort $\kappa\kappa$, as explained in Chapter 4.)

We round off the discussion with a brief word about infinitary logics. We saw earlier that 'there are infinitely many' is $\mathcal{L}_{\omega_1\omega_1}$-definable by an ω-length conjunction of inequations prefaced by an existential quantifier over ω-many variables. More generally, $\mathcal{L}_{\infty\infty}$ defines 'there are κ many' for any infinite cardinal κ in the same manner, by prefacing a κ-length conjunction of inequations with an existential quantifier over the κ-many variables contained in the inequations.

In contrast, no logic $\mathcal{L}_{\kappa\lambda}$ can define all cardinality quantifiers. The argument is effectively the same as that for SOL, ringing the changes. Let $\mu = \max(\kappa, \lambda)$ and consider $\mathcal{L}_{\kappa\lambda}(\mu)$, which contains μ-many predicate and individual constants and individual variables (as well as the usual FOL vocabulary). Since its set of sentences is of size μ, $\mathcal{L}_{\kappa\lambda}(\mu)$ cannot define all cardinality quantifiers (and indeed none beyond a certain point).[44] An analogous argument to the one for SOL shows that $\mathcal{L}_{\kappa\lambda}$, with any fixed number of predicate constants, individual constants, and individual variables, can define no more cardinality quantifiers than the $\mathcal{L}_{\kappa\lambda}(\mu)$-definable ones. An entirely analogous argument extends to any sublogic of FTT of sort $\kappa\lambda$. And by a further extension, it also applies to any sublogic of FTT of sort distinct from $\infty\infty$, as we sketch in the next paragraph.

As is familiar, nothing is gained by allowing $\lambda > \kappa$ in the language $\mathcal{L}_{\kappa\lambda}$, since the extra quantificational power cannot deliver sentences that could not be expressed—up to logical equivalence—in $\mathcal{L}_{\kappa\kappa}$; the extra quantifiers in such sentences would be 'inert'. A sentence of

[42] In the sense that $\langle V_I, \in \rangle$ is an initial segment of any other model of ZFC^2.
[43] Väänänen (2012) has the details.
[44] Since a set cannot be a proper class.

pure $\mathcal{L}_{\infty\lambda}$ (also known as pure-equality $\mathcal{L}_{\infty\lambda}$) holds in all structures of size greater than or equal to λ or it holds in none (Dickmann 1985, p. 318). By a similar sort of argument to that earlier in this appendix, any sentence true in all and only domains of size $\geq \mu$ is equivalent to a sentence of pure $\mathcal{L}_{\infty\lambda}$. These considerations, suitably elaborated, deal with $\mathcal{L}_{\infty\lambda}$ and $\mathcal{L}_{\kappa\infty}$. The argument sketch for a general sublogic of $FTT_{\infty\infty}$ whose sort is not $\infty\infty$ is similar. If such a logic defines the cardinality quantifier 'there exist at least μ' with sentence σ then by the same sort of argument as earlier in this appendix, σ is equivalent to a sentence σ^* of a *pure* sublogic of $FTT_{\infty\infty}$ obtained by quantifying out σ's non-logical predicates, constants, and function symbols. The resulting sentence σ^* may not be in the same sublogic of $FTT_{\infty\infty}$ but may belong to one of higher type (e.g. quantifying over predicate places turns a first-order sentence into a second-order one), although it will still be a sentence of some sublogic of $FTT_{\infty\infty}$, of the same sort as that of the logic σ belongs to. The upshot is that, as far as defining cardinality quantifiers goes, we may restrict attention to *pure* sublogics of $FTT_{\infty\infty}$. And now cardinality considerations show that any pure sublogic of $FTT_{\infty\infty}$ of sort $\kappa\lambda$ cannot define absolutely many quantifiers. In addition, similar to the earlier case of $\mathcal{L}_{\kappa\infty}$, logics of sort $\kappa\infty$ may also be disregarded, as their extra quantificational capacity is unexploitable. The final case is that of pure sublogics of sort $\infty\lambda$ which can be dealt with by an argument similar to the earlier one about $\mathcal{L}_{\infty\lambda}$.

6

Isomorphism Invariance

6.1 Introduction

Chapter 5 offered some bottom-up arguments for our claim that the one true logic is maximally infinitary. Let's now turn to the top-down argument for this same conclusion, starting with Tarski.

In his 1936 paper on logical consequence, Tarski notes that the characterization of formality is incomplete. Formal accounts of logical consequence must be supplemented with a demarcation of the logical constants. To quote him at length:

> Further research will doubtless greatly clarify the problem which interests us. Perhaps it will be possible to find important objective arguments which will enable us to justify the traditional boundary between logical and extra-logical expressions. But we also consider it to be quite possible that investigations will bring no positive results in this direction, so that we shall be compelled to regard such concepts as 'logical consequence', 'analytical statement', and 'tautology' as relative concepts which must, on each occasion, be related to a definite, although in greater or less degree arbitrary, division of terms into logical and extra-logical. The fluctuation in the common usage of the concept of consequence would—in part at least—be quite naturally reflected in such a compulsory situation.
>
> (Tarski 1936, p. 420)

Tarski later disowned the relativist idea suggested in this passage. He stressed that the logical constants are those expressions, such as 'not', 'and', and 'every', that are required to carry out correct inference in any field whatsoever and which feature in the most general laws. They are not expressions that have any particular subject matter themselves. In his (1986), Tarski put forward an invariance criterion as a precise test for demarcating the logical constants, understood as the most general pieces of vocabulary. In this paper, he defines logic as the 'science which deals with the notions invariant under the widest class of transformations', which will be 'very few notions, all of a very general character' (Tarski 1986, p. 149). We discussed Tarski's views on logic further in Chapter 1. As we'll now show, they form the basis of our most promising account of logical relations and logical constanthood.[1]

---

[1] We leave to others the (pre)history of the invariantist conception of logical relations and logical constants prior to Tarski. MacFarlane (2000) attributes an invariantist conception of logic to Kant. Based on citations from the *Begriffsschrift*, Sher (2013, pp. 173–4) does the same for Frege, excluded by MacFarlane. As far as we know, Mautner (1946) is the first to suggest analysing logicality in terms

One True Logic: A Monist Manifesto. Owen Griffiths and A.C. Paseau, Oxford University Press.
© Owen Griffiths and A.C. Paseau 2022. DOI: 10.1093/oso/9780198829713.003.0006

6.2 Form as schematic

We saw in Chapter 1 that Tarski was out to model an intuitive notion of *formal* logical consequence. But what did he mean by this? His first attempt at cashing out his notion of formality is in his condition (F) which, as we also saw in Chapter 1, is an expression of consequence in virtue of logical form:

> **(F)**   If, in the sentences of class K and in the sentence X, the constants — apart from purely logical constants — are replaced by any other constants (like signs being everywhere replaced by like signs), and if we denote the class of sentences thus obtained from K by K', and the sentence obtained from X by X', then the sentence X' must be true provided only that all sentences of the class K' are true.
> (Tarski 1936, p. 415)

He ultimately accepts (F) as merely a *necessary* condition for logical consequence. MacFarlane argues that Tarski's commitment to (F) demonstrates that he thought logical consequence was formal in the sense that it is *schematic*:

> To say that logic is formal is often to say that it concerns itself with inference *patterns* or *schemata* ('forms') whose instances are all correct inferences, no matter what the instantiating 'matter'.   (MacFarlane 2000, pp. 36–7)

Similarly, Dutilh Novaes (2005) argues that Tarski belongs to the tradition of taking consequence to be formal in the sense of schematic.

Tarski realized, however, that understanding the formal as schematic is not sufficient to demarcate logical consequence and truth. This is because:

> Underlying the whole construction is the division of the language discussed into logical and extralogical. The division is certainly not quite arbitrary. If, for example, we were to include among the extralogical signs the implication sign, or the universal quantifier, then the definition of the concept of consequence would lead to results which obviously contradict ordinary usage. On the other hand, no objective grounds are known to me which permit us to draw a sharp boundary between the two groups of terms.   (Tarski 1936, pp. 418–9)

In other words, logical consequence is formal in the sense of schematic, but the account remains underspecified until a principled demarcation of the logical

---

of invariance. The related study of generalized quantifiers begins with Mostowski (1957), followed by Lindström (1966); see Chapter 2 of Peters and Westerståhl (2006) for a historical overview of these developments.

constants has been given. It is underspecified because it relies, as Tarski notes, on a division of vocabulary into logical and nonlogical. If we are to find an intuitive notion of formality that the model-theoretic definition can hope to capture, therefore, this discussion of schemata must be supplemented by an account of the logical constants. Tarski later offers a precise formulation of this demarcation in terms of isomorphism invariance. Before considering that, we look at other understandings of formality in order to see the motivation for such a demarcation.

6.2.1 MacFarlane on formality

MacFarlane (2000, p. 64) distinguishes three different sorts of demarcation we could provide of the logical vocabulary to supplement the thought that logic is schematic:

1-formal The norms of logic are *constitutive* of concept use *as such* (as opposed to a particular kind of concept use). 1-formal laws are the norms to which any conceptual activity—asserting, inferring, supposing, judging, and so on—must be held responsible.

2-formal The characteristic logical notions and laws are indifferent to the particular identities of different objects. 2-formal notions and laws treat each object the same (whether it is a cow, a peach, a shadow, or a number).

3-formal Logic abstracts entirely from the semantic content or 'matter' of concepts—it considers thought in abstraction from its relation to the world and is therefore entirely free of substantive presuppositions.

MacFarlane does not claim that these three senses of formality are exhaustive (see 2000, p. 50). Indeed, Dutilh Novaes (2011) argues that MacFarlane's three senses are insufficient and that there are at least eight senses in which logic is formal! Both writers, however, accept that there has been a strong tradition, especially in the twentieth century, of taking logic to be formal in some sense.

6.2.2 Form as normative

Logic is 1-formal, for MacFarlane, just if logical laws 'are the norms to which any conceptual activity—asserting, inferring supposing, judging, and so on—must be held responsible' (2000, p. 64). He provides two key examples of philosophers who held that logic is formal in the sense of providing norms for thought. The first is

Kant, and MacFarlane (2000, p. 53) offers the following quotes in support of this view: the laws of logic are called the 'necessary laws of the understanding and of reason in general, or what is one and the same, of the mere form of thought as such' (Kant 1800, p. 13), and are referred to as 'those [laws] without which no use of the understanding would be possible at all' (1800, p. 12).

The second example is Frege:

> the basic propositions on which arithmetic is based cannot apply merely to a limited area whose peculiarities they express in the way in which the axioms of geometry express the peculiarities of what is spatial; rather, these basic propositions must extend to everything that can be thought. And surely we are justified in ascribing such extremely general propositions to logic. I shall now deduce several conclusions from this logical or formal nature of arithmetic.
>
> (Frege 1885, p. 112)

Kant therefore links the *formal* to 'laws' of understanding, which suggests that logic serves a normative role. The quote from Frege uses the word 'formal', but does not link this to the idea of the normativity of logic. It is undeniable, however, that Frege did believe that logic served a normative role; as he wrote, 'it falls on logic to discern the laws of truth ... From laws of truth there follow prescriptions about asserting, thinking, judging, inferring' (Frege 1918, p. 351).

MacFarlane (2000, §3.1) provides many more examples of philosophers who have taken the formality of logic to be what he calls 1-formality. We shall not consider any further examples, however, as 1-formality is not the sort of understanding of formality that we could usefully capture with the model-theoretic conception. This is because normativity cannot serve as a criterion of demarcation of logic, even if we set aside our insistence that logic is about implication rather than inference (see the Prologue).[2] Frege is clear in the above quote, for example, that logical consequence and truth are conceptually prior to normativity: 'From laws of truth *there follow* prescriptions' (1918, p. 351, our italics).

Prescriptions about how to reason may therefore follow from the definition of logical consequence, but they cannot be said to be constitutive of logical consequence, since the norms that we would be attempting to model would be *logical* norms: they would be norms about the ways in which we should reason logically. But a definition of logical consequence determines the ways in which we should reason. As such, to say that normativity is constitutive of logical consequence is circular, because logical norms are determined by logical consequence, but we would be defining logical consequence in terms of logical norms.

---

[2] As we saw in Chapter 3, Beall and Restall (2006) would disagree.

6.2.3 Form as topic-neutral

The best example of a logician understanding logic as 2-formal is Tarski:

> since we are concerned here with the concept of logical, i.e., formal, consequence, and thus with a relation which is to be uniquely determined by the form of the sentences between which it holds, this relation cannot be influenced in any way by empirical knowledge, and in particular by knowledge of the objects to which the sentence X or the sentences of the class K refer. The consequence relation cannot be affected by replacing the designations of the objects referred to in these sentences by the designations of any other objects.   (Tarski 1936, pp. 414–5)

Here, Tarski explicitly claims that his definition of logical consequence is intended to model an intuitive notion of *formal* logical consequence, and goes on to characterize this formality as indifference to 'the designations of the objects referred to'. In MacFarlane's taxonomy, this is a clear example of 2-formality.

We have seen that Tarski's account of formality needs to be supplemented with a demarcation of the logical constants, which he provides in his posthumous 'What are the logical notions?' (1986), a lecture first delivered in 1966. There are hints, however, as to how he earlier conceived of this distinction in publications such as his *Introduction to Logic and the Methodology of the Deductive Sciences* (1937). Here, he characterizes the logical constants as 'terms of a much more general character occurring in most of the statements of arithmetic, terms which are met constantly both in considerations of everyday life and in every field of science' (1937, p. 18). The concern of *logic*, he writes, 'is to establish the precise meaning of such terms and lay down the most general laws in which these terms are involved' (1937, p. 18).

Tarski's thought is that the logical constants are distinguished by—in modern terminology—their *topic-neutrality*: logical constants are required to carry out 'inferences in any field whatsoever'. They do not concern any particular subject matter such as numbers, but rather apply to any topic.

To see what topic-neutrality amounts to, it will be helpful to distinguish two thoughts that are common in the literature:

Discrimination   A concept is topic-neutral iff it is not able to discriminate between different individuals.

Universality   A concept is topic-neutral iff it is applicable to reasoning about any domain whatsoever.

The two notions come apart most obviously in the context of mathematics. Consider the concepts *number* and *set*. Plausibly, they are not topic-neutral in the sense of Discrimination, since they apply to some objects but not others. *Number* applies only to numbers and *set* only to sets. Nevertheless, it is plausible that these mathematical examples *are* topic-neutral according to Universality. For example, we can count any objects we like, and we can form sets of any objects we like, as long as there is some bound on their rank.[3]

Unlike mathematical concepts, many concepts usually taken as logical are topic-neutral according to both definitions. The concepts expressed by the universal quantifier and the identity predicate, for example, do not discriminate between objects and apply to reasoning in all domains.

In his (1986) paper on logical notions, Tarski combines his view of logic as 2-formal with his belief in the topic-neutrality of the logical constants. His primary concern in this paper is the demarcation of the subject matter of various disciplines, and he uses ideas drawn from Klein's *Erlangen Program* for the classification of different geometries in terms of transformations.

Klein's thought was that geometries can be distinguished by the transformations under which their central operations are *invariant*. Size, for instance, is invariant under Euclidean transformations in Euclidean space. If we apply a Euclidean transformation to, say, a planar triangle, the size (area or length) of the original triangle and that of the triangle generated by the transformation is the same. Size is not maintained, however, when we consider affine transformations, which include scalings. The more transformations we apply to a universe of discourse, the fewer the properties of objects that are invariant under those transformations. Thus Tarski:

> Now suppose we continue this idea, and consider the still wider class of *all* one-one transformations of the space, or universe of discourse, or 'world' onto itself. What will be the science which deals with the notions invariant under the widest class of transformations? Here we will have very few notions, all of a very general character. I suggest that they are the logical notions, that we call a notion 'logical' if it is invariant under all possible one-one transformations of the world onto itself.   (Tarski 1986, p. 149)

His idea is that, since logic is concerned with, as he previously put it, 'terms of a much more general character' (1937, p. 18), its notions should be preserved under the most general class of transformations.[4] Intuitively, his thought is that an operation will be logical just if it is entirely indifferent to the individual characteristics of objects. In this way, it captures the formality of logical vocabulary,

---

[3] MacFarlane (2000) has more on the distinction between these sorts of topic-neutrality, especially with respect to mathematics.

[4] Mautner (1946) independently devised a similar demarcation of logical vocabulary. See also Mostowski (1957) and Lindström (1966).

understood as topic-neutral or general applicability, and may be sharpened further into MacFarlane's notion of 2-formality. Furthermore, Tarski believed that an invariantist criterion for logical nature was vindicated by a technical result proved by himself and Lindenbaum (1935):

> Every notion defined in *Principia Mathematica*, and for that matter in any other familiar system of logic, is invariant under every one-one transformation of the 'world' or 'universe of discourse' onto itself.   (Tarski 1986, p. 150)

We shall come back to this result later in the chapter and also prove an analogue of its converse.

6.2.4 Form as abstraction

Logic is 3-formal, for MacFarlane, just if it abstracts entirely from the semantic content of interpreted sentences. At first, this may appear similar to 2-formality and its indifference to the particular identities of objects. There is a crucial difference, however: 2-formality involves abstraction away from the particular identities of some objects (those that fail an invariantist test). 3-formality, on the other hand, is stronger than 2-formality by abstracting away from all semantic content. So we abstract away from not only the nonlogical concepts, but also the logical ones. In MacFarlane's words, '[t]he word "entirely" is essential here: a logic that abstracts from the contents of *some* concepts ("specific" ones, like "horse" or "red"), but not from the content of others ("general" or logical ones, like existence, identity, and conjunction), does not count as 3-formal' (2000, p. 60).

MacFarlane argues that 3-formality can be found at various points in Kant, and later in the work of Carnap (see his 2000, §3.3 for the references, and Dutilh Novaes 2011, §2.2.2 for further investigation of the notion of 3-formality). We shall proceed by following MacFarlane's 2-formality, which is the most plausible intuitive understanding for Tarski's criterion to capture.

6.3 Permutation and isomorphism invariance

To explain isomorphism invariance, which cashes out the notion of 2-formality, we first need to explain permutation invariance. A *permutation* π is a function from a set S to S with two properties. The first is that no two distinct elements have the same value under π, that is, π is one-to-one (injective). The second is that everything in S is mapped to by some element or other under π, that is, π is onto (surjective). Informally, a permutation 'shuffles' the objects it acts upon. For example, ACB and BCA can be thought of as permutations of ABC, and the function $f(x) = x + 1$ on the integers is a permutation since it is both one-to-one and onto.[5] But the function $g(x) = x + 1$ is *not* a permutation of the natural numbers,[6] nor does $h(x) = x^2$ permute the integers.[7] Another way of putting this is that a permutation is a bijection from a set to itself.

A function on a set of objects can easily be extended to a function on sets whose members are these objects. The permutation π which maps A to B, B to C, and C to A may for example be extended to a function on the eight sets whose members are some, none, or all of A, B, or C. This extension of π maps $\{A\}$ to $\{B\}$, $\{B\}$ to $\{C\}$, $\{C\}$ to $\{A\}$, $\{A, B\}$ to $\{B, C\}$, $\{A, C\}$ to $\{B, A\}$, $\{B, C\}$ to $\{C, A\}$, $\{A, B, C\}$ to $\{B, C, A\}$ (i.e. to itself) and the empty set to the empty set. And this extension may itself be further extended by defining it on sets of sets over these same objects, sets of sets of sets, and so on. Under such an extension, $\{\{A\}\}$ would be mapped to $\{\{B\}\}$, $\{\{B\}\}$ to $\{\{C\}\}$, $\{\{\{A\}\}\}$ to $\{\{\{B\}\}\}$, and so on. In a standard abuse of notation, we use the same symbol for the original permutation as well as all these extensions, so that for example $\pi(\{\{A\}\}) = \{\{B\}\}$. More formally, the extension map is defined by the stipulation that $\pi(S) = \{\pi(s) : s \in S\}$ for any set S whose transitive closure does not contain any non-sets other than A, B, or C. Notice that some sets, such as $\{A, B, C\}$, are mapped to themselves by π, and that others, such as $\{A, B\}$, are not.

Armed with this, we can now easily explain what it is for a relation, considered as a set, to be permutation invariant. Take a domain of individuals D, for example all the people currently alive. This domain is of type i, the domain of (unary) properties over this domain is of type (i), the domain of (unary) second-order properties over these is of type $((i))$, and so on. To illustrate the idea behind permutation invariance, consider a relation R of type $((i), (i), (i))$ over the domain D; R is thus a ternary relation of unary properties of type (i), that is, properties of individuals. For example, R might be the relation associated with the intersection operation of its type, so that R consists of all triples $\langle P_1, P_2, P_3 \rangle$, each P_i being a subset of D and $P_3 = P_1 \cap P_2$. To say that R is insensitive to the identity of elements in the domain D is to say that the extension of R does not vary when we permute D. In other words, for any underlying permutation π of D, its extension to relations (considered as sets) leaves the relation R fixed: $\pi(R) = R$. Another way of putting is that if $\langle P_1, P_2, P_3 \rangle$ is in R then so is $\langle \pi(P_1), \pi(P_2), \pi(P_3) \rangle$, where $\pi(P_i) = \{\pi(x) : x \in P_i\}$, $P_i \subseteq D$ for $i = 1, 2, 3$. If R is functional, as it is when R is the intersection operation, then R is an operation, so that we may write $R(P_1, P_2) = P_3$. In that case, another way to express R's permutation invariance would be

$$R(\pi(P_1), \pi(P_2)) = \pi(R(P_1, P_2))$$

[5] One-to-one because if $f(x) = f(y)$ then $x = y$. Onto because if y is an integer then $y = f(y - 1)$.
[6] g is not onto because the natural number 0 is not equal to any natural number plus 1.
[7] h is not one-to-one because e.g. $(-1)^2 = 1^2 = 1$.

for all P_1 and P_2, since $R(\pi(P_1), \pi(P_2))$ and $\pi(R(P_1, P_2))$ are in that case both equal to $\pi(P_3)$, where $P_3 = R(P_1, P_2)$.

Contrast, say, the relation *being a profound thinker*, of type (i), over the four-membered domain consisting of Aristotle, Hypatia, Donald Trump, and Paris Hilton. We trust we won't ruffle too many feathers by taking its extension to be the doubleton set of Aristotle and Hypatia. Yet the image of {Aristotle, Hypatia} under any permutation that takes Greeks to Americans is {Donald Trump, Paris Hilton}, a set distinct from {Aristotle, Hypatia}. As the relation's extension on the permuted domain is distinct from its extension's image under the permutation, it fails to be permutation invariant over this domain.

Tarski considered permutations from a model's domain to itself. Following Sher (1991), we may generalize his idea by considering bijections from one domain to any other domain of the same size. Restricted to relations of type at most two up from the type of individuals, the resulting account of logicality has become known as the *Tarski–Sher thesis* (or *criterion*). Bijections of this kind can then be extended to a map from the finite-type hierarchy over the first domain to the finite-type hierarchy over the second. Here we go beyond Tarski and Sher, who stop at two types up from that of individuals. Since we see no reason to do so, we generalize their account along this dimension.

For a simple example, consider a first domain whose two individuals are A_1 and A_2 and a second domain whose two individuals are B_1 and B_2. The bijection that takes A_1 to B_1 and A_2 to B_2 may be extended to a map from the finite-type hierarchy over the first domain to that over the second domain. (Recall the Chapter 4 discussion of finite-type hierarchies.) Under this map, the image of the identity relation of type (i, i) over the first domain is the identity relation of type (i, i) over the second, since the set $\{\langle A_1, A_1 \rangle, \langle A_2, A_2 \rangle\}$ is mapped to $\{\langle B_1, B_1 \rangle, \langle B_2, B_2 \rangle\}$. We may think of this extended map as an isomorphism, since it bijectively maps entities of a particular type over the first domain to entities of that very same type.[8] If we denote the finite-type hierarchy over the first domain by $\mathcal{M}_1$ and that over the second by $\mathcal{M}_2$, an isomorphism from the first to the second may be written as $\pi : \mathcal{M}_1 \cong \mathcal{M}_2$.

As a further illustration, take the relation *being human* over the self-same four-membered domain made up of Aristotle, Hypatia, Donald Trump, and Paris Hilton. Since all four elements of the domain are human, the relation's extension is the universal set over that domain and is invariant under any permutation:

$$\pi(\{A, H, DT, PH\}) = \{A, H, DT, PH\}.$$

But now suppose we consider a second domain, made up of four objects none of which is human. Then if π is a bijection from the first domain to the second, its

---

[8] We could equally have callled it an extended bijection.

extension to an isomorphism over the entire finite-type hierarchy will *not* preserve the extension of *being human*. Over the first domain, the extension is the set consisting of all four elements, whose image under π is the set consisting of all four elements of the second domain; this image is distinct from the extension of *being human* over the second domain, which is the empty set.

Before proceeding further, we stipulate that a *relation* will henceforth mean a relation-in-extension, and we will often silently take its extension to be a set. For example, the relation of identity, denoted by the identity predicate 'is equal to', is individuated by its extension over all domains; in particular, over the domain $\{A, B\}$ this is the set $\{\langle A, A, \rangle, \langle B, B\rangle\}$. The relation denoted by, say, 'is a cube' and 'is a cube or a regular heptahedron' is thus one and the same—because there are no regular heptahedra—even if these expressions have different intensions (meanings).[9]

The preceding discussion gives rise to two questions. First, which relations are isomorphism invariant? That is, taking isomorphism invariance as a necessary and sufficient condition for logicality: which relations are logical? We answer this question in §6.4, and justify the answer in §6.6. As we shall see, the L∞G∞S Hypothesis in fact only requires one half of this claim: that all isomorphism-invariant relations are logical.

Second, assuming we know which *relations* are logical, which *expressions* are logical? This question is much more difficult than the first, and we won't answer it in this book. Fortunately, however we don't have to in order to establish the L∞G∞S Hypothesis. As we shall see in §6.4, all we need from our discussion is a necessary condition on the one true logic, and we can extract such a condition from isomorphism invariance without giving anything like a complete account of the logical constants. We come back to the vexing, but for us much less pressing, question of the logical constants in Chapter 11.

6.4 A necessary condition on OTL

Chapter 5 provided a 'bottom-up' argument for the thesis that the one true logic's sort is maximally infinitary. We now provide a 'top-down' argument to the same effect.

The argument's first component is (our version of) McGee's theorem. This theorem relates the relations invariant under isomorphism (in the sense of §6.3) to those expressible by formulas in the language of pure FTT$_{\infty\infty}$ (for which, see §4.6). It states that a relation is isomorphism invariant just when, for any (non-zero)

<hr/>

[9] The heptahedron example is from Gómez-Torrente (2002, p. 18), who also provides other examples, e.g. 'is a male widow'. In this work and others, such as his (2021), Gómez-Torrente has criticized the invariantist approach Sher and we favour, preferring instead a 'pragmatic' account of the logical constants. See Part III for discussion of criticisms of invariantism.

cardinal κ, its extension on a domain of size κ is expressible by a formula of $\text{FTT}_{\infty\infty}$. Or, for a more formal version, that if R is a relation over finite-type models then

for all finite relational-type models $\mathcal{M}_1, \mathcal{M}_2$ and isomorphisms
$$\pi : \mathcal{M}_1 \cong \mathcal{M}_2, \pi(R_{\mathcal{M}_1}) = R_{\mathcal{M}_2}$$
$\Updownarrow$
for each non-zero cardinal κ, there is a formula ϕ_κ^R of pure $\text{FTT}_{\infty\infty}$ such that
$$\text{if } |D_{\mathcal{M}}| = \kappa, R_{\mathcal{M}} = (\phi_\kappa^R)^{\mathcal{M}}$$

To unpack the terminology, the term '$R_{\mathcal{M}}$' refers to the relation R as applied to the model $\mathcal{M}$; $R_{\mathcal{M}}$ is what we grasp when we grasp what R's extension on $\mathcal{M}$ is. For example, if the relation R is the unary property *being a horse*, then in a domain $D_{\mathcal{M}}$ consisting of Bucephalus, RinTinTin, and Donald Trump, its extension would be the singleton of Bucephalus. This extension is determined not by interpreting a sentence of a formal language but by the nature of the relation of being a horse and by the elements of the domain $D_{\mathcal{M}} = \{$Bucephalus, RinTinTin, Donald Trump$\}$. The relation *being a horse*, of type (i), is clearly not isomorphism invariant; for example, its extension on three-individual domains can consist of 0, 1, 2, or 3 of these three individuals, depending on how many of them are horses. An example of a relation that is isomorphism invariant is the property *being a unary property of individuals*, of type $((i))$, which all properties of type (i) instantiate.

The sentence ϕ_κ^R in contrast is a sentence of a formal language, viz. the pure language of $\text{FTT}_{\infty\infty}$, and its interpretation in a model $\mathcal{M}$ is $(\phi_\kappa^R)^{\mathcal{M}}$. The notation $\pi : \mathcal{M}_1 \cong \mathcal{M}_2$ states that the map π from model $\mathcal{M}_1$ to $\mathcal{M}_2$ is an isomorphism. The theorem thus equates the informal notion $R_{\mathcal{M}}$ with the formal notions $(\phi_\kappa^R)^{\mathcal{M}}$ for each non-zero cardinal κ $(1, 2, 3, \ldots, n, \ldots, \aleph_0, \aleph_1, \ldots)$.

The second component is to find some way to use the account of isomorphism-invariant relations to pin down the nature of the logical constants. That is, to lay down a condition on logical constanthood. The ideal, of course, would be to find a necessary and sufficient condition. Plausibly, whether an expression counts as logical should supervene on its meaning. Suppose for example that the meaning of an expression is its intension (in some sense). Then the question becomes: which intensions count as logical? Invariance criteria could then be applied to intensions, and indeed have been. Refining and extending Tarski, Gila Sher for example takes a logical constant as any expression that in our terminology denotes an isomorphism-invariant relation *and* satisfies some additional conditions (see Chapter 11). However, convincing necessary and sufficient conditions for logical constanthood are hard to come by, and invariantists have struggled to find them.

Logical relations, which are worldly entities, are a different kettle of fish. Many invariantists believe that a relation is logical iff it is isomorphism invariant. Although even here there is disagreement. And even if this much is agreed, one major question is how to extend the account to modal and other operators.

In brief, few invariantists propose that *expressions* (linguistic entities) are logical iff they denote isomorphism-invariant relations. Isomorphism invariance as a criterion for the logicality of relations is more attractive. Evidently, isomorphism invariance about relations is quite distinct from isomorphism invariance about expressions, and it will be important in what follows to keep this distinction firmly in mind.

That every single invariantist criterion about logical constants has proved, at best, controversial is stressed by several authors, for example in MacFarlane (2015, §5). Here we will not endorse any such account, as we will not need to; our $L\infty G\infty S$ Hypothesis does not require it. What we will assume is that a relation—a worldly entity, individuated extensionally—is logical if it is isomorphism invariant; note that we write 'if' and not 'iff'. We extract a *necessary* condition on the one true logic from this extensional version of isomorphism invariance and use it to support our claim that this logic is maximally infinitary. We return to the correct account of logical constants, of secondary importance to us, in Chapter 11.

In proceeding this way, we shall talk about expressions and their meanings, but avoid commitment to any particular theoretical account of meaning. It would be easy to supplement our account with, say, an intensional semantics along the lines of Carnap/Montague/Kaplan/Lewis. We won't do so because we see no real need for it, other than the occasional dotting of i's and crossing of t's. Naturally, any account of meaning we could help ourselves to should not be model-theoretic, since model theory already builds in the distinction between logical and non-logical constants. But that wouldn't prevent us from helping ourselves to an account of meaning, in which for example different expressions are assigned different intensions and thereby extensions. In any case, as explained, there will be no need to assume any theoretical account of meaning; we can instead proceed intuitively. We assume only that we have a sufficient grasp of the intended interpretation of the language. In §6.3, we used the example of the domain consisting of individuals A_1 and A_2. We can perfectly well grasp that the extension of 'is equal to' as applied to individuals over this domain is the set $\{\langle A_1, A_1 \rangle, \langle A_2, A_2 \rangle\}$ even in the absence of a theoretical account of meaning.

In this spirit, suppose we are given a relation R (meaning a relation in extension) over a collection of models of a logic $\mathcal{L}$, of some particular type. If $\mathcal{L}$ is a candidate for the one true logic, it must satisfy the following condition:

$$\text{Relation } R \text{ is isomorphism invariant over } \mathcal{L}\text{-models}$$
$$\Downarrow$$
$$\mathcal{L} \text{ must contain an expression } e \text{ whose extension is equal to } R \text{ in all } \mathcal{L}\text{-models}$$

Another way to put this is as a necessary condition on the one true logic: OTL must contain logical constants for all the isomorphism-invariant relations over its models. As mentioned earlier, many invariantists believe that the logical relations

are precisely the isomorphism-invariant ones. Here we require only one direction
of this conditional, namely that isomorphism-invariant relations are logical. So
another way to put our point is that OTL must contain logical constants for all the
logical relations over its models; each logical relation must be denoted by some
OTL-constant. The nature of these constants' intensions is immaterial to our thesis,
but is an interesting further question, discussed in Chapter 11.

The informal idea behind the condition and its import for the first-order quantifiers' logical status should be clear enough. Consider a domain containing one or
more things. However we permute the entities in this domain, any instantiated
properties will be mapped to instantiated properties. Moreover, this continues
to hold if we map the domain bijectively into another domain. Likewise for
the universal quantifier: however we permute or biject entities in a domain, the
universal set of individuals is mapped to the universal set of individuals. Roughly,
this is why the usual first-order quantifiers pass the test. Indeed, many more
quantifiers than these pass. An example is 'there are finitely many': however we
permute or biject finitely many things, there will still be finitely many of them.
Another example is 'there are infinitely many': however we permute or biject
infinitely many things, there will still be infinitely many of them. And so on.

Let's put this a little more formally. Take the property of type $((i))$ instantiated
by all and only instantiated properties of type (i). In model $\mathcal{M}_1$, the property's
extension consists of all and only unary properties P over the domain instantiated
by at least one individual in the domain; similarly for model $\mathcal{M}_2$. If π is an
isomorphism from $\mathcal{M}_1$ to $\mathcal{M}_2$ (a bijection from the first's domain to the second's
domain extended to all types over these domains) then P is instantiated in $\mathcal{M}_1$
iff $\pi(P)$ is instantiated in $\mathcal{M}_2$. Thus the original property *being an instantiated
property of type (i)* is isomorphism invariant. Happily, FOL contains an expression
for this isomorphism-invariant property since it contains the existential quantifier
$\exists$, which is of type $((i))$. Similarly for the universal quantifier $\forall$, which denotes
the isomorphism-invariant relation of type $((i))$ instantiated by the universal
property of type (i). A similar conclusion holds for the identity relation of type
(i, i), which is readily seen to be isomorphism invariant. Once again, FOL contains
the expression $=$ whose denotation is precisely this relation over its models. The
formal arguments for these conclusions are so straightforward that we omit them.

Consider next the property of type $((i))$ instantiated by all and only properties
of type (i) instantiated by infinitely many things. FOL does not contain an expression whose denotation is this isomorphism-invariant property, and so fails this
necessary condition on the one true logic. In other words, FOL cannot be OTL. The
point generalizes to other infinite cardinalities. Consider the property of type $((i))$
instantiated by all and only properties of type (i) that are instantiated by at least
κ individuals. This property is obviously isomorphism invariant, and the formal
argument in §6.6 will confirm that it is; so any candidate one true logic must
contain an expression for it. The point generalizes to other types.

It is also easy to see that OTL must allow existential quantification over any number of argument places. The simplest case is that of the existential quantifier of type $((i))$, which can be thought of as the property *being an instantiated property of type (i)*. In a domain over two individuals, A and B, there are three instantiated properties of type (i), $\{A\}$, $\{B\}$, and $\{A, B\}$, so this existential quantifier's extension may be taken as $\{\{A\}, \{B\}, \{A, B\}\}$. Any model isomorphic to this first one has two individuals, A' and B', so that for any bijection from the first domain to the second (which maps A to A' and B to B' or alternatively A to B' and B to A'), the image of $\{\{A\}, \{B\}, \{A, B\}\}$ under the bijection's extension is $\{\{A'\}, \{B'\}, \{A', B'\}\}$, which is the extension of the existential quantifier of type $((i))$ in the second model. This informal argument may be generalized to an existential quantifier over any number of argument places, and of any type in the finite-type hierarchy, as well as to universal quantifiers. The formal version of the argument will follow in §6.6.

Boolean operations may be accommodated into our extensional framework by considering them as functions from finite tuples of truth-values (T and F) to these values themselves. Generalized Boolean operations—involving κ-ary conjunction and disjunction—may be accommodated in the same way. This sort of 'type-theoretic' framework, in which everything is a function, is often the preferred way in linguistics, especially linguistics in the Montagovian tradition (Montague 1974). Under it, compositionality is an instance of functional application (as Frege urged), and the treatment of quantifiers and Boolean connectives is a pleasingly unified whole.

We decided against going down this route for presentational reasons. Casting Part II's formal framework in terms of functions to truth-values would complicate the technical exposition and put off some readers. It is simpler to consider the extension of a property to be a subset of a domain than a function from the domain to the set of truth-values (elements instantiating the property corresponding to those mapped to 'true'). But once one is comfortable with the formal material, it is obvious enough how to recast everything to incorporate truth-values, and we may assume that done. It is then equally obvious to see that generalized Boolean operations drop out as isomorphism invariant, since their values do not depend on the model's particular features. For example, $\aleph_0$-conjunction is isomorphism invariant because it expresses the same operation on the countably many truth-values $v_1, \ldots, v_n, \ldots$ in any model $\mathcal{M}_1$ and thus any model $\mathcal{M}_2$ isomorphic to $\mathcal{M}_1$. $\aleph_0$-conjunction is simply the operation which takes $v_1, \ldots, v_n, \ldots$ to T iff all the v_i are T.

Incidentally, McGee (1996) reaches the same conclusion by a different route. He offers an interpretation of '$\wedge$' that treats it as set-theoretic intersection, '$\vee$' as union, '$\neg$' as complement, and so on. The connectives thus interpreted then pass the invariance test because their associated operations are isomorphism invariant.

This completes the outline of the top-down argument for the L∞G∞S Hypothesis. If OTL is a sublogic of FTT$_{\infty\infty}$, it must be of sort $\infty\infty$; more strongly, any sublogic of both OTL and $\text{FTT}_{\infty\infty}$ must be of sort $\infty\infty$. §6.5 contains some comments on this result, and §6.6 provides the argument's formal backbone.

6.5 Miscellaneous comments

We round off the discussion of the top-down argument with some comments. First, the argument has many other implications for the one true logic than the ones just drawn. It implies that the one true logic contains not just universal or existential quantification over any number of variables and Boolean operations of any arity, but all kinds of other devices as well. In particular, it contains a whole host of other generalized quantifiers and branching quantifiers. These, the subject of much work in linguistics, are not our principal concern here. Our main concern is to establish the L∞G∞S Hypothesis, a thesis radical enough in today's philosophical climate.

Second, we only treat expressions denoting an element of the finite-type hierarchy, as explained earlier. This encompasses a good deal of cleaned-up natural language. Manifestly, however, it does not include all of it. To extend, amend, or revise a result along these lines to all declarative discourse would involve an extensive exercise in natural-language semantics. Indeed, it would require no less than a complete and final version of that subject. It would, for example, involve treating not just names of functions or individuals, predicates and various determiners, but also the rest of these determiners, adverbs, plurals, conjunctions, operators, and so on. There are arguments and controversies about each of these, which is one reason we think we're a long way from knowing what more precisely the one true logic is. But again, given our purposes, there is no reason for us to enter this particular fray. However natural-language semantics pans out, we know that the one true logic will at least extend FOL. It will allow for existential and universal quantification, and contain truth-functional connectives. Given our arguments in this chapter and the previous one, it follows that the one true logic is at least as strong as $\text{FOL}_{\infty\infty}$ (i.e. $\mathcal{L}_{\infty\infty}$), that is, wildly stronger than any foundational logic generally recognized as such by most contemporary logicians. We can safely reach that conclusion, we believe, without knowing what any of natural-language semantics, or modal logic, or truth theory, etc. look like in their final form.

The third point is that, as presented here, isomorphism invariance is sensitive to the choice of semantics. Our basic thought is that the argument will go through on any natural semantics for FTT. We do not pretend to have a definition of what such a semantics must look like, but it should at least contain domains acting as extensions for every type. We can build in truth-values as well, but there are other ways of getting the same result without invoking them (as already mentioned, one option is McGee's approach in his 1996).

Perhaps the best way to frame the argument is as follows. We have offered a top-down argument for the L∞G∞S Hypothesis based on a particular kind of semantics. If you are swayed by what we say and accept a different but not too different semantics, rework the arguments in your preferred setting and you will end up in a similar place, or so we expect. If you buy a very different sort of semantics for your logic (or perhaps none at all), you might end up in a very different place. In other words, our top-down argument is addressed to those who believe in a certain kind of model-theoretic semantics, which is the great majority of logicians. To get everyone on board, we would have to have a knockdown argument up our sleeve that this or something like this is the correct semantics, an argument which we do not possess. Of course, the usual 'judge them by their fruits' argument on behalf of our chosen semantics is a strong consideration in its favour. Many logicians would, and do, find it compelling.

A fourth point concerns the quantifier 'there are absolutely many', meaning more than κ for any cardinal κ. Our version of isomorphism invariance has not underwritten this quantifier's logicality.[10] One could see this as an artefact of using set-sized domains in the semantics of $FTT_{\infty\infty}$ we gave in §4.7 (ditto for the restriction of the semantics to $FOL_{\infty\infty}$ or $SOL_{\infty\infty}$). If our metatheory had allowed class domains, we could have argued as above that 'there are absolutely many' is logical; more generally, the isomorphism-invariant relations one can refer to in such a class theory would go beyond those expressible in a sublogic of $FTT_{\infty\infty}$.[11] Relatedly, one could try to argue, in the same vein as in Chapter 5, that the argument with premises 'There are at least κ planets' for every cardinal κ and conclusion 'There is an absolute infinity of planets' is logically valid.

As far as the top-down argument is concerned, the reason we chose a set-theoretic metatheory is for simplicity and familiarity. Another approach might be to adopt a higher-order metatheory, a third to adopt a metatheory that exploits plural discourse. But given their unfamiliarity, we felt that these alternatives would impose too great a burden on the reader, for relatively little gain; and that discussing the various metatheories' pros and cons would be rebarbative and distracting. As for the bottom-up argument, we are interested in arguments framable in extensions of English. If you think there are no languages of absolutely

---

[10] There may be other arguments that this quantifier is logical. Clearly, lacking an argument that a quantifier is logical is not the same as having an argument that it is not logical.

[11] Tim Button and Sean Walsh seem to think it a criticism of Tarski–Sher that it vindicates 'two-to-the-proper-class-many' functions as logical: '…there are two-to-the-power-proper-class-many bijection-invariant quantifiers which behave like existential quantifiers on domains of *certain* cardinalities (those mapped to 0), and like universal quantification on domains of certain *other* cardinalities (those mapped to 1).' (2018, p. 406). But here they are clearly using a metatheory stronger than ours, indeed a metatheory apparently stronger than any of the usual ones, in order to allow for two layers of classes so as to define 'two-to-the-proper-class-many' functions. (ZFC, for example, cannot define 'two-to-the-proper-class-many' functions on the cardinals, nor can standard class versions of iterative set theory, such as von Neumann Bernays Gödel or Morse Kelley, in which there is only one layer of classes, and so there is no way of raising to a class power.) Other things being equal, a model theory that admits classes, or *a fortiori* layers of classes, will discern more isomorphism-invariant relations and functions than one that does not.

infinite size, you will think that for every infinite cardinal κ, there is a language with κ terms, sentences, etc. but that no language has κ terms and sentences for every κ. (To infer that there is a language of size at least κ for every κ from the fact that for every κ there is a language of size at least κ would be to commit a familiar quantifier fallacy.) So there is no argument with premises 'There are at least κ planets' for every cardinal κ and conclusion 'There is an absolute infinity of planets'. Alternatively, if you think there are languages of absolutely infinite size, you are likely to see the argument as valid.

In short, we stand four-square behind the thesis that the one true logic's sort is maximally infinitary. Whether 'there are absolutely many' is also a logical quantifier and must on that account feature in the one true logic is a nicety that depends on other commitments. We leave this further question for future work.

An analogy helps clarify what is at stake. Suppose a theist comes up with an argument that an omnipotent being exists. That being can do everything that can be done: not just move mountains but create and destroy the universe, and more. These consequences of omnipotence can be appreciated by anyone who understands the term. But can an omnipotent being also create a stone they cannot lift? That is a harder question, and hinges on relatively subtle matters. It is also of secondary importance. Establishing the existence of an omnipotent being would constitute a major philosophical advance, even in the absence of a complete account of what omnipotence comes to. (Which is why most theists, even most reflective ones, are not particularly concerned by the paradox of the stone.) Similarly, it is an important thesis in the philosophy of logic that the one true logic exists and is maximally infinitary, even in the absence of a complete account of what maximal infinity comes to. This means, at the very least, that conjoining any number of formulas and quantifying over any number of argument places are logical operations and must be represented by logical constants in the one true logic. Whether it means that the quantifier 'there are absolutely many' should join logical company with all the 'there are κ-many' quantifiers is a further issue. We believe we have offered strong evidence for the $L\infty G\infty S$ Hypothesis, that the one true logic's sort is maximally infinitary. We have drawn some important consequences from it; but we don't claim to have teased them all out.

The fifth point harks back to a question implicitly raised in Chapter 1. Our account of logical consequence in this book is based on formality, spelt out in a particular way. We saw in Chapter 1 that many (but not all) authors take necessary truth-preservation to be a feature of logical consequence, and moreover that several of them make it a cornerstone of their account. Although we do not repudiate necessary truth-preservation, we do not make it our starting point either. There's a sense, though, in which we end up getting the best of both worlds, because formality yields necessary truth-preservation. The necessity of consequence in fact follows from an isomorphism-invariance-based model-theoretic account, coupled with the assumption that any possible domain is isomorphic to some model or other. For in that case, if sentence s is not necessary because it is false in some possible domain (under reinterpretation, holding its logical terms fixed), it must also be false in an isomorphic actual model. Contraposing, it follows that all logical truths are necessary; and similarly for logical consequence. As this point has been made clearly by others,[12] we will not elaborate it further here. Observe, as Gómez-Torrente (2021, p. 32) astutely does, that it is not the formal account on its own that delivers this feature of consequence. It is the formal account combined with the thesis that any possible interpretation is isomorphic to an actual model (e.g. a set-theoretic model).

The sixth and final point is that the argument for (our version of) McGee's Theorem, set out in §6.6, may be given in an informal metatheory. It could be formalized in, for example, first-order ZFC. It could also be formalized in a set theory cast in a much stronger logic (e.g. $FOL_{\infty\infty}$ or $FTT_{\infty\infty}$) so long as this set theory permits standard constructions on domains of any cardinality (power set, separation, etc.). This observation is philosophically important. For if, say, $SOL_{\infty\infty}$ is the one true logic, then the correct set theory must be cast in this logic. If we give ZFC cast in $SOL_{\infty\infty}$ the name $ZFC^{SOL_{\infty\infty}}$, this theory must allow the arguments in this chapter to go through in order to vindicate $SOL_{\infty\infty}$ as the one true logic. Without spelling out the details, it's clear that adopting a stronger theory of models than first-order ZFC, for example $ZFC^{SOL_{\infty\infty}}$ (whatever exactly that is), would also underwrite this chapter's argument. If first-order ZFC suffices for carrying out the argument, so do its extensions. We also note in passing that the necessary condition above on OTL depends sensitively on which models are in play; we return to this point in §7.2.

6.6 McGee's Theorem

This section, the last and longest of the present chapter, presents a central technical result, McGee's Theorem, stated at the start of §6.5 and repeated below in §6.6.1. Informally, our version of this theorem says that a relation is isomorphism invariant just when its extension on any domain of size κ is expressible by a formula of pure $FTT_{\infty\infty}$. McGee's Theorem forms an important premise in our argument for the L∞G∞S Hypothesis, as we saw. For its importance lies in linking isomorphism invariance to infinitary logic. Readers familiar with McGee's Theorem, or who wish to follow the philosophical thread without interruption, may head straight to Chapter 7.

McGee (1996) aims to answer the question of what the logical operations are, given isomorphism invariance. In that paper, McGee is interested only in the extensional characterization of these operations: he takes a connective to be logical

[12] Sher (2008, 2013) and Gómez-Torrente (2021).

if and only if, on each domain, the operation on the domain described by the connective is invariant under permutations (McGee 1996, p. 570).[13] He defines a notion of permutation invariance on a domain and then proves that an operation on a domain is invariant in his sense iff it is described by some formula of $\mathcal{L}_{\infty\infty}$ (= FOL$_{\infty\infty}$) on all domains of that cardinality. Such a formula need not be a formula of pure $\mathcal{L}_{\infty\infty}$ but may contain non-logical constants of type (i). McGee then sketches two extensions of this result. The first is to expressions of higher type; the second is to operations that are not just invariant in his sense under permutations but under bijection too.

Earlier drafts of this book contained a full exposition of McGee's 1996 paper. This material turned out be heavy going, because McGee's argument is intricate and his notation dense. In the end, we thought better of including it and will instead present a much simpler argument that cuts through some of the complexity of McGee's own proof. In particular, McGee makes heavy use of variable assignments, whereas we avoid them, for a simpler treatment. Another difference is that McGee aims to capture logical operations rather than logical relations. This difference is merely one of emphasis: operations may be thought of as relations with an extra argument place. For example, conjunction may be thought of as the two-place intersection operation or as the three-place relation, whose third element is the intersection of the first two. A third difference is that McGee is principally concerned with operations expressible in the language $\mathcal{L}_{\infty\infty}$ (= FOL$_{\infty\infty}$), rather than relations/operations in the language FTT$_{\infty\infty}$, as we are. He *seems* to consider only models over finite types with non-logical constants of type (i), although as the discussion of his first extension is extremely brief, it is hard to conclude this with certainty. Be that as it may, it is not clear what would justify a restriction to type (i), philosophically speaking. Our approach will be to consider all finite types. Nonetheless, to be clear, McGee (1996) is a major advance, to which much of our discussion in this section is owed.

6.6.1 Isomorphism invariance

The purpose of the present section is to prove each of the two directions of the following biconditional.

for all finite relational-type models $\mathcal{M}_1, \mathcal{M}_2$ and bijections
$$\pi : \mathcal{M}_1 \cong \mathcal{M}_2, \pi(R_{\mathcal{M}_1}) = R_{\mathcal{M}_2}$$
$$\Updownarrow$$
for each non-zero cardinal κ, there is a formula ϕ_κ^R of pure FTT$_{\infty\infty}$ such that
$$\text{if } |D_{\mathcal{M}}| = \kappa, R_{\mathcal{M}} = (\phi_\kappa^R)^{\mathcal{M}}$$

---

[13] An important antecedent to both McGee's theorem and our version of it is Lindenbaum and Tarski (1935).

This theorem is inspired by McGee (1996) and Lindenbaum and Tarski (1935). Given these results, it is a short mathematical step to the theorem we prove here. In particular, our approach owes much to the first and second extension of McGee's Theorem.[14] We do, however, believe that our way of putting things expresses some of the ideas in the existing literature more clearly.

Two points should be noted before we embark on the proof. The first is that individuals are covered by the theorem, since they may be considered to be relations of type i. Second, as mentioned earlier, our discussion excludes sentential connectives. These qualify as isomorphism-invariant if we move to a type-theoretic framework in which there are primitive types e and t, for entities and truth-values respectively, and compounds (x, y) from available types. A property of individuals is on this approach of type (e, t), as it maps entities to truth-values; binary connectives are of type $((t, t), t)$, and so on. The approach trivially deems propositional connectives of arbitrary arity isomorphism-invariant, since their type involves t only, hence they are not affected by mappings on the domain of entities.[15] This easy extension will be implicit throughout.

6.6.2 Relations

We now prove the first half of the equivalence by showing that a formula of the pure language of $\text{FTT}_{\infty\infty}$ expresses a relation that is isomorphism invariant. (Recall $\text{FTT}_{\infty\infty}$'s definition from Chapter 4; 'pure' here means that no non-logical vocabulary is included.)[16] That is to say, if for all κ, some formula ϕ_κ^R of pure $\text{FTT}_{\infty\infty}$ expresses the relation $R_{\mathcal{M}}$ on all $\mathcal{M}$ of size κ,[17] then $\pi(R_{\mathcal{M}_1}) = R_{\mathcal{M}_2}$ for any two isomorphic models $\mathcal{M}_1, \mathcal{M}_2$ related by the isomorphism π, that is, $\pi : \mathcal{M}_1 \cong \mathcal{M}_2$. As earlier, if $\pi : D_{\mathcal{M}_1} \to D_{\mathcal{M}_2}$ is a bijection, we define $\pi(R_{\mathcal{M}_1}^{(\tau_1, \ldots, \tau_n)}) = \{\langle \pi(R_{\mathcal{M}_1}^{\tau_1}), \ldots, \pi(R_{\mathcal{M}_1}^{\tau_n}) \rangle : \langle R_{\mathcal{M}_1}^{\tau_1}, \ldots, R_{\mathcal{M}_1}^{\tau_n} \rangle \in R_{\mathcal{M}_1}^{(\tau_1, \ldots, \tau_n)}\}$, for all relations $R_{\mathcal{M}_1}^{(\tau_1, \ldots, \tau_n)}$ of non-individual type.

Thus suppose that $\pi : \mathcal{M}_1 \to \mathcal{M}_2$ is an isomorphism between two finite-type models $\mathcal{M}_1$ and $\mathcal{M}_2$, each of cardinality κ. We must show that if R is expressed by the formula ϕ_κ^R on all models of size κ, with $R_{\mathcal{M}_1} = (\phi_\kappa^R)^{\mathcal{M}_1}$ and $R_{\mathcal{M}_2} = (\phi_\kappa^R)^{\mathcal{M}_2}$ and $\pi : \mathcal{M}_1 \cong \mathcal{M}_2$, then $\pi(R_{\mathcal{M}_1}) = R_{\mathcal{M}_2}$; equivalently, that $\pi((\phi_\kappa^R)^{\mathcal{M}_1}) = (\phi_\kappa^R)^{\mathcal{M}_2}$. We prove this by showing by induction that for *any* formula ϕ of pure $\text{FTT}_{\infty\infty}$, $\pi(\phi^{\mathcal{M}_1}) = \phi^{\mathcal{M}_2}$.

---

[14] Our argument for our claim about $\text{FTT}_{\infty\infty}$ is based on the construction briefly mentioned on page 575 in McGee (1996). And our argument for the converse is a variant of the argument in Lindenbaum and Tarski (1935).

[15] van Benthem (1989) contains much more on this approach and perspective.

[16] As noted, the usual distinction between logical and non-logical vocabulary can be drawn extensionally, by providing lists of each. We are *not* presupposing a criterion of logical constanthood.

[17] That is, $|D_{\mathcal{M}}| = \kappa$. For brevity, we often say that a model is of size κ when its domain is of that size.

Some notational points: for clarity, in the rest of this section we use '$\simeq$' for the object-language identity symbol and reserve '$=$' for the metalanguage identity symbol. To reduce clutter, we usually drop the subscript '$\mathcal{M}$' in the expression '$R_{\mathcal{M}}$' when it's obvious what the model in question is. Finally, recall the §4.7 definition of $E_{\mathcal{M}}^{\tau}$ as the set of all elements of type τ in model $\mathcal{M}$.

On to the inductive argument. The first two cases are the induction's base cases, in which ϕ is $X^{(\tau_1,\ldots,\tau_n)}(X^{\tau_1},\ldots,X^{\tau_n})$ or $X_i^{\tau} \simeq X_j^{\tau}$. For the first case, we must prove that:

$$\pi((X^{(\tau_1,\ldots,\tau_n)}(X^{\tau_1},\ldots,X^{\tau_n}))^{\mathcal{M}_1}) = (X^{(\tau_1,\ldots,\tau_n)}(X^{\tau_1},\ldots,X^{\tau_n}))^{\mathcal{M}_2}$$

The proof is immediate:

$$\pi((X^{(\tau_1,\ldots,\tau_n)}(X^{\tau_1},\ldots,X^{\tau_n}))^{\mathcal{M}_1})$$
$$= \pi(\{\langle R^{\tau_1},\ldots,R^{\tau_n},R^{(\tau_1,\ldots,\tau_n)}\rangle \in E_{\mathcal{M}_1}^{\tau_1} \times \ldots \times E_{\mathcal{M}_1}^{\tau_n} \times E_{\mathcal{M}_1}^{(\tau_1,\ldots,\tau_n)} : \langle R^{\tau_1},\ldots,R^{\tau_n}\rangle \in$$
$$R^{(\tau_1,\ldots,\tau_n)}\})$$
$$= \{\langle \pi(R^{\tau_1}),\ldots,\pi(R^{\tau_n}),\pi(R^{(\tau_1,\ldots,\tau_n)})\rangle \in E_{\mathcal{M}_2}^{\tau_1} \times \ldots \times E_{\mathcal{M}_2}^{\tau_n} \times E_{\mathcal{M}_2}^{(\tau_1,\ldots,\tau_n)} : \langle R^{\tau_1},\ldots,R^{\tau_n}\rangle \in$$
$$R^{(\tau_1,\ldots,\tau_n)}\}$$
$$= \{\langle R^{\tau_1},\ldots,R^{\tau_n},R^{(\tau_1,\ldots,\tau_n)}\rangle \in E_{\mathcal{M}_2}^{\tau_1} \times \ldots \times E_{\mathcal{M}_2}^{\tau_n} \times E_{\mathcal{M}_2}^{(\tau_1,\ldots,\tau_n)} : \langle R^{\tau_1},\ldots,R^{\tau_n}\rangle \in$$
$$R^{(\tau_1,\ldots,\tau_n)}\}$$
$$= (X^{(\tau_1,\ldots,\tau_n)}(X^{\tau_1},\ldots,X^{\tau_n}))^{\mathcal{M}_2},$$

the penultimate equation following from the fact that π is an isomorphism. Similarly,

$$\pi((X_i^{\tau} \simeq X_j^{\tau})^{\mathcal{M}_1})$$
$$= \pi(\{\langle R_i^{\tau},R_j^{\tau}\rangle \in E_{\mathcal{M}_1}^{\tau} \times E_{\mathcal{M}_1}^{\tau} : R_i^{\tau} = R_j^{\tau}\})$$
$$= \{\langle \pi(R_i^{\tau}),\pi(R_j^{\tau})\rangle \in E_{\mathcal{M}_2}^{\tau} \times E_{\mathcal{M}_2}^{\tau} : R_i^{\tau} = R_j^{\tau}\}$$
$$= \{\langle R_i^{\tau},R_j^{\tau}\rangle \in E_{\mathcal{M}_2}^{\tau} \times E_{\mathcal{M}_2}^{\tau} : R_i^{\tau} = R_j^{\tau}\}$$
$$= (X_i^{\tau} \simeq X_j^{\tau})^{\mathcal{M}_2}.$$

The penultimate equation once more follows from the fact that π is an isomorphism. This proves the induction's two and only base cases.

For the induction step, suppose Φ takes the form $\bigvee\{\phi_i : i < \kappa\}$, where each ϕ_i is a formula in the pure language of $\mathsf{FTT}_{\infty\infty}$. If the result holds for each ϕ_i then using the induction hypothesis (IH):

$$\pi((\bigvee\{\phi_i : i < \kappa\})^{\mathcal{M}_1})$$
$$= \pi(\bigcup\{\phi_i^{\mathcal{M}_1} : i < \kappa\})$$
$$= \bigcup\{\pi(\phi_i^{\mathcal{M}_1}) : i < \kappa\}$$
$$\underset{IH}{=} \bigcup\{\phi_i^{\mathcal{M}_2} : i < \kappa\}$$
$$= (\bigvee\{\phi_i : i < \kappa\})^{\mathcal{M}_2}$$

For the negation case, note first that if ϕ is a pure $\mathsf{FTT}_{\infty\infty}$-formula with λ free variables then $\phi^{\mathcal{M}}$ consists of all the λ-tuples $\langle R_0^{\tau_0}, \ldots, R_\alpha^{\tau_\alpha}, \ldots \rangle$ satisfying ϕ.[18] Since the αth element of such a tuple is an element of $E_{\mathcal{M}}^{\tau_\alpha}$, the interpretation of ϕ in $\mathcal{M}$, $\phi^{\mathcal{M}}$, is a subset of $\bigtimes_{\alpha<\lambda} E_{\mathcal{M}}^{\tau_\alpha}$.[19] It follows that $(\neg\phi)^{\mathcal{M}} = \bigtimes_{\alpha<\lambda} E_{\mathcal{M}}^{\tau_\alpha} \backslash \phi^{\mathcal{M}}$.[20] Thus:

$$
\begin{aligned}
&\pi\big((\neg\phi)^{\mathcal{M}_1}\big) \\
&= \pi\big(\bigtimes_{\alpha<\lambda} E_{\mathcal{M}_1}^{\tau_\alpha} \backslash \phi^{\mathcal{M}_1}\big) \\
&= \bigtimes_{\alpha<\lambda} E_{\mathcal{M}_2}^{\tau_\alpha} \backslash \pi(\phi^{\mathcal{M}_1}) \\
&\underset{IH}{=} \bigtimes_{\alpha<\lambda} E_{\mathcal{M}_2}^{\tau_\alpha} \backslash \phi^{\mathcal{M}_2} \\
&= (\neg\phi)^{\mathcal{M}_2}
\end{aligned}
$$

Finally, for the existential quantifier, suppose ϕ takes the form $\exists\{X_i^{\tau_i} : i < \kappa\}\psi$. We may assume for notational simplicity that, relative to a well-ordering of the variables, the variables $\{X_i^{\tau_i} : i < \kappa\}$ immediately extend the free variables in ϕ (i.e. the free variables in ψ that are not bound in ϕ). For example, suppose that in a well-ordering of variables the first three variables are x, y, and z (in that order); then ϕ may take the form $\exists z(x \simeq y \vee y \simeq z)$, since its free variables are x and y and its existential quantifier-variable pair is $\exists z$, z being the next variable after x and y in the well-ordering. We say that a relational tuple $\vec{R}$ is *apt* for a formula if its elements' type corresponds to the types of the formula's free variables, that is, the tuple is of the kind that can satisfy the formula relative to a variable assignment (which does not of course mean that it *does* satisfy the formula). For example, $\vec{R} = \langle R_1^i, R_3^{(i,(i))} \rangle$ is apt for the formula $X^i \simeq X^{(i,(i))}$ (and of course $R_1^i \neq R_3^{(i,(i))}$ since the two relations are of different type). We use the symbol $\frown$ for concatenation of tuples, so that $\vec{R} \frown \vec{S}$ is the tuple $\vec{R}$ followed by the tuple $\vec{S}$; e.g. $\langle R_1, R_2 \rangle \frown \langle R_3, R_4 \rangle = \langle R_1, R_2, R_3, R_4 \rangle$.

With this notation in place, the argument for the existential case $\phi = \exists\{X_i^{\tau_i} : i < \kappa\}\psi$ is straightforward:

$$
\begin{aligned}
&\pi\big((\exists\{X_i^{\tau_i} : i < \kappa\}\psi)^{\mathcal{M}_1}\big) \\
&= \pi\big(\{\phi\text{-apt } \vec{R} : \vec{R} \frown \vec{S} \in \psi^{\mathcal{M}_1} \text{ for some } \vec{S} \text{ such that } \vec{R} \frown \vec{S} \text{ is } \psi\text{-apt}\}\big) \\
&= \{\phi\text{-apt } \pi(\vec{R}) : \vec{R} \frown \vec{S} \in \psi^{\mathcal{M}_1} \text{ for some } \vec{S} \text{ such that } \vec{R} \frown \vec{S} \text{ is } \psi\text{-apt}\} \\
&\underset{IH}{=} \{\phi\text{-apt } \pi(\vec{R}) : \pi(\vec{R} \frown \vec{S}) \in \psi^{\mathcal{M}_2} \text{ for some } \vec{S} \text{ such that } \pi(\vec{R} \frown \vec{S}) \text{ is } \psi\text{-apt}\}
\end{aligned}
$$

---

[18] We assume here some arbitrary well-ordering of the variables, or at least of the free variables in ϕ.

[19] The symbol '$\bigtimes$' denotes the Cartesian product, so that for example $\bigtimes_{\alpha<2} F_\alpha = F_0 \times F_1$. The symbol '\' denotes set-theoretic complementation.

[20] We have written the right-hand side in this way rather than as $(\bigtimes_{\alpha<\lambda} E_{\mathcal{M}}^{\tau_\alpha})\backslash\phi^{\mathcal{M}}$, in order to reduce clutter.

$$= \{\phi\text{-apt } \pi(\vec{R}) : \pi(\vec{R}){\frown}\pi(\vec{S}) \in \psi^{\mathcal{M}_2} \text{ for some } \vec{S} \text{ such that } \pi(\vec{R}){\frown}\pi(\vec{S}) \text{ is } \psi\text{-apt}\}$$
$$= \{\phi\text{-apt } \vec{R} : \vec{R}{\frown}\vec{S} \in \psi^{\mathcal{M}_2} \text{ for some } \vec{S} \text{ such that } \vec{R}{\frown}\vec{S} \text{ is } \psi\text{-apt}\}$$
$$= (\exists\{X_i^{\tau_i} : i < \kappa\}\psi)^{\mathcal{M}_2}$$

The more general case, in which the variables $\{X_i^{\tau_i} : i < \kappa\}$ do not (or need not) immediately extend the free variables in ϕ, is a notational variant of the argument just given.

We have proved that relations expressed by pure $\mathrm{FTT}_{\infty\infty}$-formulas are isomorphism invariant. So for any relation R, if for all cardinals κ, there is a formula ϕ_κ^R of pure $\mathrm{FTT}_{\infty\infty}$ such that if $|D_{\mathcal{M}}| = \kappa, R_{\mathcal{M}} = (\phi_\kappa^R)^{\mathcal{M}}$, then R is isomorphism invariant.

6.6.3 Caveats

It is important to be clear about what the result in §6.6.2 implies and what it does not. First, it does not follow from the fact that ϕ expresses R in *some* model $\mathcal{M}$ that R is isomorphism-invariant. For example, the relation *being a horse* is expressed by the formula $X^i \simeq X^i$ in the model with domain {Marengo}, the singleton whose only member is Napoleon's war mount. But the property of being a horse is clearly not isomorphism invariant; in a non-equine domain its extension is empty. Similarly, the fact that there is a particular cardinal κ and formula ϕ_κ^R such that $R_{\mathcal{M}} = (\phi_\kappa^R)^{\mathcal{M}}$ for all models of size κ does not imply that R is isomorphism-invariant. For example, consider the property *is a property of type (i) which has no more than 2 instances or is a property of type (i) whose instances are all horses*. This property is of type $((i))$ and is expressed by the formula $X^{(i)} \simeq X^{(i)}$ in all domains of size no greater than 2. But it isn't isomorphism invariant. Its extension in a domain of size 3 all of whose elements are horses is the set of all properties of type (i); in contrast, its extension in a domain of size 3 with one non-equine element is not the set of all properties of type (i).

From this direction of the theorem it follows at once that no formula of *full* $\mathrm{FTT}_{\infty\infty}$ can play the role of ϕ_κ^R, unless it is equivalent to a formula of pure $\mathrm{FTT}_{\infty\infty}$ over domains of cardinality κ. This applies for example to formulas such as Fx, where F is a non-logical monadic predicate constant of type (i) and x is a variable of type i. This is *not* to say that Fx cannot express an isomorphism-invariant relation in *some* model (in which constants such as F are given an interpretation/extension). For example, in a model $\mathcal{M}$ in which the extension of F is $D_{\mathcal{M}}$, $\mathcal{M}$'s domain, the elements of $D_{\mathcal{M}}$ that satisfy Fx are all the elements of $D_{\mathcal{M}}$. In this model, Fx expresses an isomorphism-invariant relation, expressible in all models by the formula $x \simeq x$, where x is a variable of type i. Thus it may be that F shares its extension with that of an isomorphism-invariant relation in some model, although it does not do so in all models.

6.6.4 Isomorphism-invariant relations

We now deal with the biconditional's other direction. Suppose that R, a relation of finite-type relations, is isomorphism invariant, that is, that for all finite-type models $\mathcal{M}_1, \mathcal{M}_2$ such that $\pi : \mathcal{M}_1 \cong \mathcal{M}_2$, $\pi(R_{\mathcal{M}_1}) = R_{\mathcal{M}_2}$. We prove that if κ is a (non-zero) cardinal then there is a formula ϕ_κ^R of pure $\mathcal{L}_{\infty\infty}^{\text{FTT}}$ whose interpretation on any model of size κ is $R_{\mathcal{M}}$, that is, if $|D_{\mathcal{M}}| = \kappa$ then $R_{\mathcal{M}} = (\phi_\kappa^R)^{\mathcal{M}}$.

The idea behind the construction of ϕ_κ^R is simple. Let $\mathcal{M}$ be a finite-type model of size κ and $R_{\mathcal{M}}$ be R's extension in this model. The first step is to well-order all the elements of each type (any well-order will do). Call the αth relation of non-individual type $\tau = (\tau_1, \ldots, \tau_n)$ under this well-order R_α^τ. For each such R_α^τ, define a sentence $\text{Diag}(R_\alpha^\tau)$ of pure $\mathcal{L}_{\infty\infty}^{\text{FT}}$ that is the conjunction over all n-tuples $\langle R_{\alpha_1}^{\tau_1}, \ldots, R_{\alpha_n}^{\tau_n} \rangle$ of

$$X_\alpha^\tau(X_{\alpha_1}^{\tau_1}, \ldots, X_{\alpha_n}^{\tau_n}) \text{ if } \langle R_{\alpha_1}^{\tau_1}, \ldots, R_{\alpha_n}^{\tau_n} \rangle \in R_\alpha^\tau;$$
$$\neg X_\alpha^\tau(X_{\alpha_1}^{\tau_1}, \ldots, X_{\alpha_n}^{\tau_n}) \text{ if } \langle R_{\alpha_1}^{\tau_1}, \ldots, R_{\alpha_n}^{\tau_n} \rangle \notin R_\alpha^\tau.$$

Model theorists will recognize $\text{Diag}(R_\alpha^\tau)$ as the free-variable version of the diagram of R_α^τ, where this diagram is understood as a long conjunction. Clearly, $\text{Diag}(R_\alpha^\tau)$ will be infinitary iff $D_{\mathcal{M}}$ is infinite.

Given $\text{Diag}(R_\alpha^\tau)$, next define $\text{Diag}_\downarrow(R_\alpha^\tau)$ to be 'the' formula $\text{Diag}(R_\alpha^\tau)$ conjoined with $\text{Diag}(R_\beta^{\tau^*})$ for any β and $\tau^* \neq \tau$ such that the free variable $X_\beta^{\tau^*}$ appears in $\text{Diag}(R_\alpha^\tau)$, and repeat this procedure for the resulting sentence and all subsequent ones. The procedure ends after finitely many steps since all relations are of finite type, and are therefore well-founded. More formally, $\text{Diag}_\downarrow(R_\alpha^\tau)$ is any sentence S of pure $\text{FTT}_{\infty\infty}$ that (i) contains $\text{Diag}(R_\alpha^\tau)$ as a conjunct, and (ii) is such that if the free variable $X_\beta^{\tau^*}$ appears in S, where $\tau^* \neq i$, then $\text{Diag}(R_\beta^{\tau^*})$ is a conjunct in S; and (iii) S is a (proper or improper) subsentence, up to the order of conjuncts, of any sentence with properties (i) and (ii). In the typical case, $\text{Diag}_\downarrow(R_\alpha^\tau)$ in fact defines a set of sentences that are identical up to the order of their conjuncts rather than a single sentence. But since we may always choose a sentence from this set if necessary, and any such sentence is as good as any other, we speak of 'the' sentence $\text{Diag}_\downarrow(R_\alpha^\tau)$.[21]

The penultimate step is to conjoin $\text{Diag}_\downarrow(R_\alpha^\tau)$ with the formula stating that all the i-variables are distinct and the universal i-quantifier stating that any i-variable must be equal to one of those appearing free in $\text{Diag}_\downarrow(R_\alpha^\tau)$. Thus if there are κ variables in $\text{Diag}_\downarrow(R_\alpha^\tau)$, because there are κ elements in $\mathcal{M}$'s domain, the resulting formula is

---

[21] Given that every type is well-ordered, it would be easy to define $\text{Diag}_\downarrow(R_\alpha^\tau)$ uniquely if desired.

$$\text{Diag}_{\downarrow}(R_{\alpha}^{\tau}) \wedge \bigwedge_{\alpha_1 < \alpha_2 < \kappa} \neg X_{\alpha_1}^i \simeq X_{\alpha_2}^i \wedge \forall Y^i (\bigvee_{\alpha < \kappa} Y^i \simeq X_{\alpha}^i),$$

where Y^i is a variable of type i distinct from all the i-variables X_{α}^i. (Variables ranging over individuals are usually written as lower-case letters; here we write them as X^i, Y^i etc. for notational uniformity.)

The final step is to preface the resulting formula with an existential quantifier that quantifies over each free variable appearing in the formula other than X_{α}^{τ}. The resulting formula is then ϕ_{κ}^R. Note that ϕ_{κ}^R's only free variable is X_{α}^{τ}, since all the others have been bound in this last step, or in the case of Y^i, the penultimate step.

We illustrate our definition with a simple example. Suppose $D_{\mathcal{M}}$ has two elements, a_0 and a_1, so that $\kappa = 2$. We well-order all the elements of each type over this domain. Suppose that under this ordering $R_0^{(i)}$ is the empty relation (of type (i), as the superscript indicates), $R_1^{(i)}$ holds of a_0 only, $R_2^{(i)}$ holds of a_1 only, and $R_3^{(i)}$ holds of both a_0 and a_1. The relation $R_6^{((i))}$ is the $((i))$-relation which holds of $R_1^{(i)}$ and $R_2^{(i)}$ only; intuitively, it's the property 'is a property of individuals had by exactly one of the two individuals'. For this example:[22]

$\text{Diag}(R_0^{(i)})$ is $\neg X_0^{(i)} X_0^i \wedge \neg X_0^{(i)} X_1^i$;

$\text{Diag}(R_1^{(i)})$ is $X_1^{(i)} X_0^i \wedge \neg X_1^{(i)} X_1^i$;

$\text{Diag}(R_2^{(i)})$ is $\neg X_2^{(i)} X_0^i \wedge X_2^{(i)} X_1^i$;

$\text{Diag}(R_3^{(i)})$ is $X_3^{(i)} X_0^i \wedge X_3^{(i)} X_1^i$;

$\text{Diag}(R_6^{((i))})$ is $\neg X_6^{((i))} X_0^{(i)} \wedge X_6^{((i))} X_1^{(i)} \wedge X_6^{((i))} X_2^{(i)} \wedge \neg X_6^{((i))} X_3^{(i)}$;

$\text{Diag}_{\downarrow}(R_6^{((i))})$ is $\text{Diag}(R_6^{((i))}) \wedge \text{Diag}(R_0^{(i)}) \wedge \text{Diag}(R_1^{(i)}) \wedge \text{Diag}(R_2^{(i)}) \wedge \text{Diag}(R_3^{(i)})$;

$\phi_2^{R_6^{((i))}}$ is $\exists X_0^i \exists X_1^i \exists X_0^{(i)} \exists X_1^{(i)} \exists X_2^{(i)} \exists X_3^{(i)} (\neg X_0^i \simeq X_1^i \wedge \forall X_2^i (X_2^i \simeq X_0^i \vee X_2^i \simeq X_1^i) \wedge \text{Diag}_{\downarrow}(R_6^{((i))}))$

In this last formula, we wrote what we earlier referred to as Y_i as X_2^i instead, to emphasize its distinctness from the other two variables of the same type, X_0^i and X_1^i. $\phi_2^{R_6^{((i))}}$ in all its unabbreviated glory is thus:

$$\exists X_0^i \exists X_1^i \exists X_0^{(i)} \exists X_1^{(i)} \exists X_2^{(i)} \exists X_3^{(i)} (\neg X_0^i \simeq X_1^i \wedge \forall X_2^i (X_2^i \simeq X_0^i \vee X_2^i \simeq X_1^i) \wedge \neg X_6^{((i))} X_0^{(i)}$$
$$\wedge X_6^{((i))} X_1^{(i)} \wedge X_6^{((i))} X_2^{(i)} \wedge \neg X_6^{((i))} X_3^{(i)} \wedge \neg X_0^{(i)} X_0^i \wedge \neg X_0^{(i)} X_1^i \wedge X_1^{(i)} X_0^i \wedge \neg X_1^{(i)} X_1^i \wedge$$
$$\neg X_2^{(i)} X_0^i \wedge X_2^{(i)} X_1^i \wedge X_3^{(i)} X_0^i \wedge X_3^{(i)} X_1^i)$$

---

[22] For existential quantification over a small finite number of variables, we use standard notation for finitary languages, e.g. we write not $\exists\{x, y\}$ but $\exists x \exists y$.

Observe that the bound variables in $\phi_2^{R_6^{((i))}}$ are $X_0^i, X_1^i, X_2^i, X_0^{(i)}, X_1^{(i)}, X_2^{(i)}$, and $X_3^{(i)}$, and that the formula's only free variable is $X_6^{((i))}$. Of course, $\phi_2^{R_6^{((i))}}$ is in this case finitary since $D_{\mathcal{M}}$ is finite (it's of size 2); when $D_{\mathcal{M}}$ is infinite, $\phi_\kappa^{R^\tau}$ will be infinitary. It should be pretty clear that the unary relation of type $((i))$ *being a unary relation of type (i) which holds of exactly one thing* is isomorphism-invariant; in our model, it is the relation $R_6^{((i))}$ expressed by the formula $\phi_2^{R_6^{((i))}}$.

Returning to the general case, it now simply remains to show that the interpretation of ϕ_κ^R in our fixed model of size κ is $R_{\mathcal{M}}$. Since ϕ_κ^R's only free variable is X_α^τ, this is tantamount to showing that $(\mathcal{M}, \sigma) \vDash \phi_\kappa^R$ when $\sigma(X_\alpha^\tau) = R_\alpha^\tau$. If we let ρ be any variable assignment that maps the free variables $X_\beta^{\tau^*}$ in $\mathrm{Diag}_\downarrow(R_\alpha^\tau)$ to $R_\beta^{\tau^*}$ for any $\tau^* \leq \tau$ and any β, then $(\mathcal{M}, \rho) \vDash \mathrm{Diag}_\downarrow(R_\alpha^\tau)$, since $\mathrm{Diag}_\downarrow(R_\alpha^\tau)$ has been defined precisely so as to capture all the 'downward' facts involving R_α^τ in the model $\mathcal{M}$. (By downward facts, we mean: what the elements of R_α^τ are, what the elements of its elements are, and so on.) As any σ such that $\sigma(X_\alpha^\tau) = R_\alpha^\tau$ differs from this ρ at most over variables that are bound in ϕ_κ^R, it follows from the satisfaction clause for the existential quantifier that $(\mathcal{M}, \sigma) \vDash \phi_\kappa^R$ for any such σ.

This completes the proof that isomorphism-invariant relations are expressed by pure $\mathrm{FTT}_{\infty\infty}$-formulas.

7

Towards the One True Logic

At the start of Part II, we distinguished two dimensions of a logic: its type and sort. The L∞G∞S Hypothesis, to which Part II is devoted, is that the one true logic's sort is $\infty\infty$ if a sublogic of $FTT_{\infty\infty}$; more strongly, that the maximal sublogic of OTL and $FTT_{\infty\infty}$ is of sort $\infty\infty$; and most strongly, albeit more vaguely, that OTL's sort is maximally infinitary in a more general sense. Chapter 5 advanced some bottom-up arguments for FOL and SOL's foundational inadequacy. We saw that these arguments generalized to logics of finitary ($\omega\omega$) sort, and that they pointed to the more general moral about OTL's highly infinitary nature. Chapter 6 then offered a more theoretical and general argument for the L∞G∞S Hypothesis. We put forward a necessary condition on OTL: it must contain a logical constant for every relation that is isomorphism invariant over OTL's models. Reworking McGee's theorem, we showed that this implies that OTL allows quantification over any number of arguments and disjunction/conjunction over any number of formulas. Chapter 6's top-down argument does not supersede, but complements and strengthens, Chapter 5's bottom-up arguments. The bottom-up and top-down arguments converge to the same conclusion, that the one true logic's sort is maximally infinitary.

Our conclusions do not pin down the logic precisely—we make no bones about that. Invariantism by itself cannot deliver the one true logic, as invariantists may have hoped. It does not tell us the logic's type; and even if it could tell us what all the logical constants are, it would not tell us how they combine.[1] But it does reveal surprising features of the one true logic and significantly reduces the range of options. This is perhaps not surprising, since a logic must answer to many criteria, not just one. With hindsight, it was perhaps naive to imagine that a many-dimensional concept such as 'logical' could be defined on the basis of a single simple criterion. Nevertheless, our invariantism cuts down on the live options significantly, and rules out all the finitary logics which between them have hitherto gained quasi-universal support. In the current climate, the idea that logic is maximally infinitary is nothing short of revolutionary.

The present chapter now ties up three loose ends. The first is the question of OTL's type. The second is our account's dependence on a theory of models. The third is what the mathematical consequences of moving to a logic of sort $\infty\infty$ are.

---

[1] Compare: intuitionists and classical first-order logicians arguably accept the same logical constants but don't accept the same logical laws.

One True Logic: A Monist Manifesto. Owen Griffiths and A.C. Paseau, Oxford University Press.
© Owen Griffiths and A.C. Paseau 2022. DOI: 10.1093/oso/9780198829713.003.0007

7.1 The logic's type

Our hope is that this book answers some important questions in the philosophy of logic. Yet we are very aware that it does not answer many others, including some of no lesser importance. In this section, we consider one important direction in which to develop our ideas.

The literature of the past few decades has tended to suggest that (in our terminology) the one true logic's *type* is at least that of SOL. We briefly review some of the arguments in favour of going beyond FOL, at least as far as SOL. Much of the material here will be familiar to the *cognoscenti*. But perhaps not all of it is, and reviewing it with our aims in mind should in any case prove instructive. Our broad aim is to sketch the argument that the one true logic's type is at least that of SOL. Coupled with our hypothesis that the one true logic is maximally infinitary, this implies that $SOL_{\infty\infty}$ is a better candidate for the one true logic than $FOL_{\infty\infty}$.

In this section, then, we recall some reasons for preferring a logic of SOL's type to one of FOL's type.[2]

(i) FOL apparently cannot handle inferences involving quantification over properties or relations. An example is 'Ann is tall, Bill is tall; *so there is something Ann and Bill both are*', which seems logically valid. SOL comfortably handles such inferences.

(ii) FOL apparently cannot handle plural expressions and arguments. A classic example is the Geach–Kaplan sentence 'Some critics admire only one another', which is expressible in SOL but can be shown to be inexpressible in FOL. Examples such as 'Some sets ....' defeat any general attempt to paraphrase plural discourse about the As as first-order singular discourse about the set of As.[3]

(iii) The semantics of second-order logic is continuous with that of first-order logic. The former supplements the latter by a small number of clauses.[4]

(iv) SOL, unlike FOL, can express generalized quantifiers such as 'Most', 'There are infinitely many', 'There are uncountably many', and the like.

(v) FOL is incapable of properly characterizing structures of central importance to mathematics. SOL, however, can.[5]

The first reason still applies even if we change the logic's sort to $\infty\infty$: $FOL_{\infty\infty}$ is incapable of adequately handling inferences involving properties and relations. The argument therefore points to the one true logic's having a type as least as strong

---

[2] Many of the observations in the next few paragraphs have been made by several authors; our references are to one or two of the *loci classici*.
[3] See Boolos (1984) for discussion.
[4] See Boolos (1975) for discussion.
[5] As well as the articles by Boolos above, see Shapiro (1991).

as that of SOL. Combined with our claims about its sort, that points to the one true logic's being at least as strong as $SOL_{\infty\infty}$.

The second reason also still applies even if we change the logic's sort to $\infty\infty$: $FOL_{\infty\infty}$ is incapable of adequately handling inferences involving plurals. As with the previous point, the argument suggests that the one true logic's type is at least SOL's. Naturally, some authors prefer to render natural-language plural quantification by means of a sui generis plural logic rather than SOL.[6] This debate is orthogonal to the ones we are mainly interested in, and fans of plural logic will at least agree that the one true logic must go beyond FOL's type, as it contains a plural logic within it. We shall bracket this proviso in what follows, on the understanding that some of the reasons for extending the one true logic's type to SOL may be reasons for going beyond the finite-type hierarchy altogether. And our view, of course, is that the L∞G∞S Hypothesis applies to plural logic just as much as to singular logic: it too must be absolutely infinite, if part of the one true logic.

We are less convinced than many authors by the third reason. But we note that if it does constitute a sound argument, it supports the move to a logic whose type is not just that of SOL's but that of FTT. After all, the standard semantics for FTT naturally extends that of SOL too; roughly, it iterates the way in which SOL's semantics extends FOL's a countable infinity of times. Combined with the arguments in Chapters 5 and 6, this points to the one true logic being at least as strong as $FTT_{\infty\infty}$.

As for the fourth reason, $FOL_{\infty\infty}$ (i.e. $\mathcal{L}_{\infty\infty}$) can express any quantifier of the form 'There are at least κ', for κ a cardinal. For example, the sentence 'There are at least κ sets' may be formalized as

$$\exists\{x_i : i < \kappa\}(\bigwedge_{i<j<\kappa} x_i \neq x_j \wedge \bigwedge_{i<\kappa} Set(x_i))$$

or, as it is often written,

$$(\exists x \restriction \kappa)(\bigwedge_{\alpha<\beta<\kappa} x_\alpha \neq x_\beta \wedge \bigwedge_{\alpha<\kappa} Set(x_\alpha))$$

This is a formula of $\mathcal{L}_{\kappa^+\kappa^+}$, since it involves κ-length quantification and κ-length conjunction. Similarly, 'There are exactly κ sets' is $\mathcal{L}_{\kappa^+\kappa^+}$-formalizable.[7] We may abbreviate the formalization of 'There are exactly κ Fs' as $\exists!\kappa F$, which is a formula of $\mathcal{L}_{\kappa^+\kappa^+}$. (This notation will be used in the appendix to this chapter.)

---

[6] An example is Oliver and Smiley (2013).

[7] For the record, by the sentence

$$(\exists x \restriction \kappa)(\bigwedge_{\alpha<\beta<\kappa} x_\alpha \neq x_\beta \wedge \bigwedge_{\alpha<\kappa} Set(x_\alpha) \wedge \forall y(Set(y) \rightarrow \bigvee_{\alpha<\kappa} y = x_\alpha))$$

Yet $\mathcal{L}_{\infty\infty}$, that is, $\text{FOL}_{\infty\infty}$, cannot formalize the quantifier 'Most'. The basic reason is that 'Most Fs are Gs' places no bounds on the cardinalities of the Fs and Gs, so that any putative $\text{FOL}_{\infty\infty}$-formalization of this sentence would, *per impossibile*, have to be of absolutely infinite length. A sketch of a more formal argument may be found in the appendix to this chapter.

In sum, we find here support for moving from the logic $\text{FOL}_{\infty\infty}$ (i.e. $\mathcal{L}_{\infty\infty}$), whose type is that of FOL, to $\text{SOL}_{\infty\infty}$, whose type is SOL's. The one true logic must have maximally infinitary sort ($\infty\infty$) if a sublogic of $\text{FTT}_{\infty\infty}$, as we have seen in the previous chapters. The evidence in the present section points to the one true logic's type being at least that of SOL, since $\text{FOL}_{\infty\infty}$ will not suffice to express the generalized quantifier 'most' (and related ones).

As for the final reason—(v) in the above list—part of a foundational logic's job is to characterize mathematical structures. Doing so underwrites an account of mathematical truth as entailment from the relevant structure-characterizing axioms. A simple and familiar example of such an account takes arithmetical truth to be defined by the conditional 'ϕ is true iff $T_{AR} \vDash \phi$', where ϕ is an arithmetical claim and T_{AR} is an appropriate set of axioms. From this point of view, the task of underwriting a precise and informative account of mathematical truth is a special case of the task of capturing logical-consequence facts.

Yet FOL is not up to this task, as is well-known. The first important example of a mathematical notion definable in second- but not first-order terms was given by Frege in the *Begriffschrift*. This was the ancestral relation 'y follows in the ϕ-series after x', which Frege defined as

$$\forall F([\forall z(\phi xz \rightarrow Fz) \wedge \forall d(Fd \rightarrow \forall z(\phi dz \rightarrow Fz))] \rightarrow Fy)$$

Many other mathematical notions for which no first-order characterization exists have been second-order defined since. Kreisel (1967) was to our knowledge the first to point out that we often adopt a first-order axiom scheme because we accept its second-order formulation:

A moment's reflection shows that the evidence of the first order axiom schema derives from the second order schema: the difference is that when one puts down the first order schema one is supposed to have convinced oneself that the specific formulae used (in particular, the logical operations) are well defined in any structure that one considers.   (Kreisel 1967, p. 148).

For example, we accept the first-order axiom scheme of Induction because we accept its second-order version that quantifies over *all* properties. Or we accept the first-order set-theoretic axioms of Separation and Replacement because we accept their second-order versions.[8] As Zermelo (1930) proved, second-order set theory is quasi-categorical, meaning that if $\mathcal{M}_1$ and $\mathcal{M}_2$ are both models of this theory, then $\mathcal{M}_1$ is isomorphic to a submodel of $\mathcal{M}_2$ or vice-versa, and each model may be specified up to isomorphism by its height (i.e. by the least cardinal it does not contain, which must be strongly inaccessible).[9]

More generally, second-order formulations often capture our understanding of the respective intended structures, in a way that the collection of instances of the first-order schemata fails to. Indeed, most of the axiomatizations of mathematical structures given at the turn of the twentieth century were second-order, precisely for this reason. The usual first-order axiomatization of arithmetic, Peano Arithmetic, for instance, its many virtues notwithstanding, has one crucial vice: it does not pin down the structure of the standard model of arithmetic. This vice is common to all first-order theories of arithmetic satisfied by the standard model: if they are satisfied by the standard model they must be satisfied by a non-standard one as well. Similarly, any first-order theory of real analysis has non-standard models. More generally, the Löwenheim-Skolem theorems show that any first-order theory with an infinite model has a model of any infinite cardinality greater than or equal to that of the theory's (first-order) language.

In contrast, as is also well known, there are categorical second-order axiomatizations of arithmetic and real analysis. The second-order axiomatization of Peano Arithmetic is given by the following three axioms:

$$\neg \exists x(Sx = 0)$$
$$\forall x \forall y(Sx = Sy \rightarrow x = y)$$
$$\forall X[(X0 \land \forall x(Xx \rightarrow XSx)) \rightarrow \forall x Xx]$$

The move to a second-order version is not the only way to go, however. There are other ways to characterize the standard model of arithmetic by transcending first-order logic. The class of all models isomorphic to the standard model of arithmetic may be characterized by adding to the axioms of first-order Peano Arithmetic the following axiom, expressible in $\mathcal{L}_{\omega_1 \omega}$ (i.e. $\mathsf{FOL}_{\omega_1 \omega}$):

$$\forall x(x = 0 \lor x = S0 \lor x = SS0 \lor \ldots)$$

More generally, Scott's Isomorphism Theorem for $\mathcal{L}_{\omega_1 \omega}$ shows that any countable first-order model may be characterized uniquely (i.e. up to isomorphism) by some

---

[8] As Boolos (1984, p. 65) puts it, Separation cries out for second-order formulation.

[9] A similar quasi-categoricity result follows for set theories with urelements by fixing the cardinality of the urelemente, and a categoricity result by fixing the cardinality of the urelemente and the height of the hierarchy.

sentence of $\mathcal{L}_{\omega_1\omega}$ (i.e. $\mathsf{FOL}_{\omega_1\omega}$).[10] *A fortiori*, it is characterizable by a sentence of $\mathcal{L}_{\infty\infty}$.

As for the field of real numbers, this too can be characterized up to isomorphism by a second-order axiomatization. The basic idea is that the ordered field of real numbers (i.e. the real numbers with addition $+$ and multiplication $\times$ and the less-than relation $<$) enjoys a uniqueness property: any other structure satisfying this structure's characterizing axioms must be isomorphic to it. More precisely, the reals are the up-to-isomorphism unique ordered field with the least-upper-bound property.[11] The ordered-field axioms are all first-order expressible and the least-upper-bound property is second-order expressible. We also note in passing, as it will be relevant for Part III, that there is a second-order sentence S_{CH} which is a second-order validity iff the Continuum Hypothesis is true.[12]

We can also characterize the real numbers uniquely in $\mathcal{L}_{\infty\infty}$. As already noted, the ordered field axioms are all first-order expressible. Using infinitary resources, we can express the completeness of the field by stating that every Cauchy sequence converges to an element of the field.[13] In fact, nothing like the full strength of $\mathcal{L}_{\infty\infty}$ is needed; $\mathcal{L}_{\omega_1\omega_1}$ will suffice.

Let us take stock: as we have seen, the natural-number and the real-number structure are each definable in $\mathcal{L}_{\infty\infty}$ (i.e. $\mathsf{FOL}_{\infty\infty}$) and SOL but not in FOL. A host of more general results abound about which structures are categorically SOL-characterizable. Further such results may be proved under the hypothesis that the universe of sets is that of all constructible sets ($V = L$). For example, under this hypothesis, we have:

---

[10] That is to say, if $\mathfrak{A}$ is a countable first-order model then there is a sentence $\phi_{\mathfrak{A}}$ of $\mathcal{L}_{\omega_1\omega}$ such that for all countable first-order models $\mathfrak{B}$, $\mathfrak{B}$ (as an $\mathcal{L}_{\omega_1\omega}$-model) satisfies $\phi_{\mathfrak{A}}$ iff $\mathfrak{B}$ is isomorphic to $\mathfrak{A}$. The result is due to Dana Scott; its proof may be found in Keisler (1971, pp. 7–8) and several textbooks on model theory.

[11] Any ordered field satisfying this so-called *Completeness Axiom* is known as a complete ordered field. The axiom states that any non-empty set with an upper bound has a least upper bound.

[12] The Continuum Hypothesis (CH) is an interpreted first-order sentence in the language of set theory often expressed as $2^{\aleph_0} = \aleph_1$. It is a special case of the Generalized Continuum Hypothesis (GCH), according to which $2^\kappa = \kappa^+$ for all infinite κ.

[13] Here is an argument sketching how to define the infinitary formula $\forall\{x_i : 1 \leq i < \omega\}$ ($x_1, x_2, \ldots x_n, \ldots$ is a Cauchy sequence $\leftrightarrow x_1, x_2, \ldots, x_n \ldots$ converges to some y); note that both the antecedent and consequent of the parenthetical biconditional are infinitary formulas. Observe first that, since the theory contains terms for 0 and 1 and formal analogues for addition and multiplication, it can define each of the rationals. It can thereby define the predicate 'is rational': something is rational iff it is identical to one of the countably many rationals (the definiens here is an infinite disjunction). We can then state that $x_1, x_2, \ldots, x_n, \ldots$ is a Cauchy sequence by saying that for any positive rational q, either all the elements of the (improper) subsequence $x_1, x_2, \ldots, x_n, \ldots$ are all within q of each other, or all the elements of the (proper) subsequence $x_2, x_3, \ldots x_n \ldots$ are, or all those of the subsequence $x_3, x_4, \ldots, x_n, \ldots$ are, and so on—a countably infinite disjunction. We may then express the fact that any Cauchy sequence converges to a point in the space by stating that if any $x_1, x_2, \ldots, x_n, \ldots$ form a Cauchy sequence then there is an element y such that for any positive rational q, either all the elements of the subsequence $x_1, x_2, \ldots, x_n, \ldots$ are within q of y, or all the elements of the subsequence $x_2, x_3, \ldots, x_n, \ldots$ are, or all those of the subsequence $x_3, x_4, \ldots, x_n, \ldots$ are, and so on—again, a countably infinite disjunction. Notice that all infinitary sentences used in these definitions are countably infinite. We expect there are more elegant ways of proceeding than the brute-force formalization just sketched.

- The second-order theory of a countable structure is categorical. (Marek-Magidor/Ajtai)
- The second-order theory of Borel structure is categorical. (H. Friedman)
- A recursively axiomatizable complete second-order theory is categorical. (Solovay)

On the other hand, there may be some structures not categorically pinned down by SOL. Consistently with ZFC there is a complete and finitely axiomatizable SOL-theory with finite vocabulary that is not categorical.[14] But if we combine SOL with arbitrarily many Boolean operations, we can capture any set-like structure. In particular, every structure of cardinality κ is categorically characterizable in $SOL_{\kappa^+ \omega}$ (Baldwin 2018, p. 175). $SOL_{\infty \omega}$, and *a fortiori* $SOL_{\infty \infty}$, thus suffices for the characterization of virtually any mathematical structure.

As for $\mathcal{L}_{\infty \infty}$ ($FOL_{\infty \infty}$), it is incapable of characterizing structures that employ the notion of an arbitrary subset. In particular, as we saw earlier, a second-order characterization of set theory reflects our epistemic commitments and allows us to pin down the universe of sets, *modulo* its height. $\mathcal{L}_{\infty \infty}$ can't match SOL in this regard. To take a notable example, any model of second-order ZFC plus the claim that there is a strongly inaccessible cardinal has a strongly inaccessible domain. But it is known—and not hard to prove—that if κ is strongly inaccessible then any $\mathcal{L}_{\kappa \kappa}$-theory whose language is of cardinality smaller than κ has a model of size smaller than κ.[15] Informally put, there is an important difference between SOL and any $\mathcal{L}_{\kappa \lambda}$. SOL can claim something about the power set of a set of arbitrary size. In contrast, there is an upper cardinal bound on the number of elements $\mathcal{L}_{\kappa \lambda}$ can enumerate. $\mathcal{L}_{\kappa \lambda}$ can only mimic claims about the power set up to a point but not beyond. Any set (as opposed to a proper class) of $\mathcal{L}_{\infty \infty}$-formulas is therefore subject to this limitation.

Similarly, topological notions with an unbounded second-order definition (in which the second-order quantifiers do not range over sets of bounded size) are known to be undefinable in $\mathcal{L}_{\infty \infty}$. Applications of the infinitary downward Löwenheim-Skolem theorem for $\mathcal{L}_{\infty \infty}$ show that many types of spaces are not $\mathcal{L}_{\kappa \lambda}$-characterizable for any κ and λ. Dickmann (1985, p. 323) offers the following list:

topological spaces;
compact spaces;

---

[14] Much of the present paragraph essentially restates p. 70 of Baldwin (2018).
[15] Magidor and Väänänen (2011, p. 89).

discrete spaces;

T_i spaces ($i = 0, \ldots, 5$);

regular, completely regular, normal, completely normal spaces;

compact and any of the preceding separation [T_i] axioms;

metrizable spaces;

Stone spaces, extremally disconnected spaces;

complete uniform spaces.

To sum up, SOL has several advantages lacked by $\mathcal{L}_{\infty\infty}$, even if the latter is a vast improvement on FOL in many regards. There is a strong case to be made, then, that the one true logic's type is at least that of SOL.[16] If we are right that the one true logic's sort is maximally infinitary, the next step is to examine whether $SOL_{\infty\infty}$ or some logic of even higher type, such as $FTT_{\infty\infty}$, is the one true logic. In light of the preceding paragraphs, it seems that if there is pressure to go beyond $SOL_{\infty\infty}$ to find the one true logic, it will not come from mathematics proper. It is perhaps more likely to come from modal logic or truth theory, fields that use operators not found in second-order logic.

7.2 Theory of models

It is important to appreciate that, whatever form it takes, invariantism cannot recommend any particular logic on its own but only a logic in combination with a theory of models. A simple example provides a stark illustration. Suppose you think only one model exists, whose domain contains exactly one thing. Perhaps the reason is that you believe God and only God exists—what we take to be other things are merely His manifestations. On this metaphysical picture, the name 'God' would be isomorphism invariant: it denotes the same entity in every domain, that is, the one and only domain. So under isomorphism invariance, God would be a logical individual, and the one true logic must contain a logical constant that names God. Like most, we reject this metaphysical picture (as would theists and atheists alike). What it illustrates, though, is that an account of the logical constants depends on what models exist.

On the standard model-theoretic approach, consequence is characterized as truth-preservation over a collection of models. In standard logical practice, one is not usually very precise about what this collection consists of. We observed in Chapter 6 that our results did not sensitively depend on our range of models, as long they included all those of, say, first-order ZFC, the usual background theory of models. We take Chapter 6's conclusions to follow on any account of models that

---

[16] *Modulo* the earlier point about plural logic vs SOL.

does not differ too much from standard accounts, for we relied on quasi-universal assumptions about models. In particular, to deduce that the one true logic's sort is $\infty\infty$, we take all the models posited by standard set theory, ZFC or some extension of it, to exist. We stressed this point in §6.5 when discussing class-based versus set-based semantics.

To put it a little more formally, suppose you start off with a theory of models, cast in a logic $\mathcal{L}_1$: call it $MT_{\mathcal{L}_1}$. Today's logicians are likely to take MT as set theory, most probably ZFC, and $\mathcal{L}_1$ as FOL; but in principle $MT_{\mathcal{L}_1}$ can be any theory of models whatsoever. We then determine which relations are invariant according to $MT_{\mathcal{L}_1}$ and thus which are logical. Now let $\mathcal{L}_2$ be the logic in which all and only these relations are expressible. Since $\mathcal{L}_2$ is then a better candidate than $\mathcal{L}_1$ for OTL (if the two are distinct), it must be fed back into the theory of models. One's new theory of models should then be $MT_{\mathcal{L}_2}$, which recasts $MT_{\mathcal{L}_1}$ in a more suitable candidate logic. And so on. It's clear, then, that invariantism does not deliver a logic as such. Rather, it stipulates that the correct model-theory and logic pair, $MT_{\mathcal{L}}$ and $\mathcal{L}$, are a fixed point of this procedure.

7.3 Mathematical consequences?

Our account's philosophical significance should be clear enough. But what, we are sometimes asked, are the mathematical consequences of accepting an infinitary logic? We don't suggest that equating OTL with a highly infinitary logic suffices in itself to resolve any deep mathematical questions. But this equation brings in its train a significant change of focus in foundational studies. It should foster greater interest in infinitary logics, and generally lead to a shift in the kinds of questions logicians investigate. Just as philosophical interest in SOL has in the past few decades spurred mathematical investigation of SOL-theories, so the philosophical case for highly infinitary logics will, we hope, revive, broaden, and deepen interest in infinitary theories.

In particular, further work is needed to better understand theories cast in logics of sort $\infty\infty$. For example, if $SOL_{\infty\infty}$ is adequate for E_c (roughly cleaned-up current natural language, including mathematics and science), it would be of great interest to investigate theories framed using this logic. In particular, set theories framed in $SOL_{\infty\infty}$ ought to be a subject of great interest, although so far as we know they remain unexplored. One could look at $FTT_{\infty\infty}$-theories too, in particular $FTT_{\infty\infty}$-set theories. The underlying principle here is that one's theory of models should be cast in one's true logic.

We do not in any way suggest that the study of weaker logics or theories employing them should be discouraged, still less abandoned. Our philosophical stance is compatible with letting a thousand logics bloom, mathematically speaking; coming to a better understanding of *any* logic is mathematically valuable.

Moreover, interesting proof-theoretic investigations of logics of infinitary sort will be inherently limited by the fact that such logics are deductively incomplete.[17] We can't see a reason to stop using first-order set theories to investigate the properties of, say, $\text{SOL}_{\infty\infty}$-theories. This is entirely analogous to the use of first-order set theories to investigate the properties of SOL-theories by philosophical logicians who regard SOL as a better candidate for the one true logic than FOL. To put it another way, as the study of SOL already makes clear, the correct foundational logic's epistemic utility may be limited. Theories cast in a sublogic of OTL will be more useful to us limited cognizers, endowed with finitary reasoning powers— unlike superhumans, supersuperhumans, and divine entities if any such exist. That a finitary logic is our best epistemic handle on the infinitary OTL is neither surprising nor an argument against OTL's foundational status. In a rough analogy, our finitary minds are for theists our only means of grasping a being whose infinity transcends our full understanding. In Johannine terms, one cannot but use human words to understand the divine Word. So it is with logic: the correct one is infinitary, although to understand it, we must resort to finitary means.[18]

Our point, then, is that if our arguments in Part II are along the right lines, natural candidates for the correct foundational logic should be further investigated. This study should in the philosophically ideal case take place in a metalanguage that uses the selfsame logic, even if in some contexts a weaker logic will do for metatheoretical investigations. We now turn to Part III, in which we defend our infinitary standpoint from objections.

Appendix: the generalized quantifier 'most'

We show that 'most' is not $\mathcal{L}_{\infty\infty}$-definable. Suppose that the (infinite) cardinal λ is such that the number of cardinals below is less than λ; for example λ might be ξ^+ where ξ is a fixed point of the aleph function, so that $\xi = \aleph_\xi$ and the number of cardinals smaller than $\lambda = \xi^+$ is ξ, which is less than λ. In that case, the formula

$$\neg(\bigvee_{\mu<\xi} \exists!_\mu F)$$

is an $\mathcal{L}_{\lambda\lambda}$-formula expressing 'There are at least λ Fs', since the disjunction following the negation is of length ξ $(< \lambda)$ and each disjunct is in $\mathcal{L}_{\mu\mu}$ for $\mu < \lambda$ (including finite μ). The relevance of this observation will be clear shortly.

Consider next the following infinitary version of the downward Löwenheim-Skolem theorem:[19]

---

[17] As is well-known, there are completeness results for infinitary logics if we allow proof systems to be infinitary too.

[18] It is important, of course, that these finitary means do not lead us astray. The simplest way to ensure that our epistemically accessible logic is a good guide to OTL is that it be a sublogic of OTL.

[19] Dickmann (1985, p. 339). In what follows, we assume all cardinals are infinite.

Let $\lambda < \kappa$ be regular cardinals such that if $\mu < \kappa$ and $\nu < \lambda$ then $\mu^\nu < \kappa$.

Then every sentence of $\mathcal{L}_{\kappa\lambda}$ which has a model has a model of size $< \kappa$.

Now assume for *reductio* that $\chi(F, G)$ is an $\mathcal{L}_{\infty\infty}$-sentence defining 'Most Fs are Gs'. We let λ be a sufficiently large regular cardinal such that $\chi(F, G)$ is expressible in $\mathcal{L}_{\lambda^+\lambda}$ and set $\kappa = \lambda^+$. As in the previous paragraph, we may take $\lambda = \xi^+$ where $\xi = \aleph_\xi$, so that $\kappa = \xi^{++}$. Now consider the sentence

$$\sigma = \chi(F, G) \wedge \exists_{\geq\lambda}(F, \neg G),$$

where $\exists_{\geq\lambda}(F, \neg G)$ formalizes 'There exist at least λ Fs that are not Gs'. As we saw in the previous paragraph (replacing F with 'F that is not G'), σ is $\mathcal{L}_{\lambda^+\lambda}$-sentence. If $\mu < \kappa = \lambda^+$ then $\mu \leq \lambda$ and if also $\nu < \lambda$ then $\nu \leq \xi$, so that $\mu^\nu \leq \lambda^\xi = (\xi^+)^\xi$. Under the Generalized Continuum Hypothesis (GCH), which states that $\kappa^+ = 2^\kappa$ for every infinite cardinal κ, $(\xi^+)^\xi = \xi^+$. Assuming GCH, then, the (infinitary) Löwenheim-Skolem theorem may now be invoked to show that σ has a model of size $< \lambda^+$, that is, of size no more than λ. But σ cannot be true in a model of size no more than λ, since it states that most Fs are Gs (first conjunct) but also that at least λ Fs are not Gs (second conjunct). It follows that the $\mathcal{L}_{\infty\infty}$-expressibility of $\chi(F, G)$ contradicts GCH; so we cannot prove in standard set theory that $\chi(F, G)$ is $\mathrm{FOL}_{\infty\infty}$-expressible.

PART III
OBJECTIONS

8

The Heterogeneity Objection

8.1 Introduction

In Part II, we presented the invariance approach to logical relations (in extension). We articulated various arguments for the thesis that the one true logic is maximally infinitary, a thesis we called the L∞G∞S Hypothesis.[1] We did not argue for a specific logic as the one true logic, but showed that any candidate for it must be maximally infinitary. In Part III, we shall consider the major objections to our view.

The main objections that have been raised to isomorphism invariance are the following four. We summarize them briefly here.[2]

1. *Heterogeneity Objection:* No natural explanation is given by the account of what constitutes the same logical operation.
2. *Overgeneration Objection:* Isomorphism invariance assimilates logic to mathematics, more specifically to set theory.
3. *Absoluteness Objection:* The set-theoretical notions involved in explaining the semantics of a maximally infinitary logic such as $\mathcal{L}_{\infty\infty}$ are not absolute.
4. *Intensional Objection:* Isomorphism invariance ignores the *meanings* of logical constants.

We dedicate this and the next three chapters to answering these objections. We tackle the first in this chapter, the second in Chapter 9, the third in Chapter 10, and the last in Chapter 11. The role of the present chapter more generally is to examine other forms of invariantism. By covering these other invariantisms, we thereby rebut the heterogeneity objection.

In Part II, we assumed that invariantism takes the form of isomorphism invariance. In this chapter, we assume invariantism but consider challenges to spelling it out as *isomorphism* invariance. We shall consider three other forms of invariantism: uniform-isomorphism invariance (§8.2); Feferman's

---

[1] For simplicity, this is usually how we'll refer to the conclusion of Part II throughout Part III. The more precise formulation, as we saw, is that the one true logic's *sort* is maximally infinitary. We also saw in Part II that this thesis comes in three different versions, of different strengths.

[2] Statements of the first three objections are set out in Feferman (2010a, p. 6), whose wording we have loosely followed. We shall not make relatively fine distinctions between isomorphism invariance as a thesis about relations extensionally conceived or as a thesis about relations intensionally conceived when the writers under discussion do not do so and/or when there is no great need to do so.

One True Logic: A Monist Manifesto. Owen Griffiths and A.C. Paseau, Oxford University Press.
© Owen Griffiths and A.C. Paseau 2022. DOI: 10.1093/oso/9780198829713.003.0008

strong-homomorphism invariantism (§8.3); and Bonnay's potential-isomorphism invariantism (§8.4). We have some sympathy with the first, much less so with the second and third.

8.2 Uniform-isomorphism invariance

As we saw earlier, the Tarski–Sher criterion applied to operations or relations of a certain type takes them to be logical iff they are isomorphism invariant, that is, their extensions are isomorphism invariant on all models of size κ, for each cardinal κ. In Chapter 6, we adopted this criterion of logical extensions and extended it to all types.[3] We then derived a necessary condition on the one true logic from it. To wit: for each logical relation, and so in particular for each isomorphism-invariant relation of any type, the one true logic must contain a logical constant whose denotation is that relation in any model.

Critics of isomorphism invariance have found this problematic. Isomorphism invariance counts as logical any miscellaneous collection of relations R_κ, one per cardinal κ, such that each R_κ is isomorphism invariant even if—and therein lies the alleged rub—the R_κs have no connection to one another. Take, for example, the relation $\forall_{reg}/\exists_{sing}$, defined as the universal quantifier on all domains of size a regular cardinal and as the existential quantifier on all domains of size a singular cardinal.[4] According to isomorphism invariance, $\forall_{reg}/\exists_{sing}$ is logical; but, critics contend, this quantifier is too heterogeneous to be a logical expression.

The sternest critic of isomorphism invariance, Solomon Feferman, has pressed this objection in several papers. As he put it, the Tarski–Sher thesis affords 'no natural explanation .... of what constitutes the *same* logical operation over arbitrary basic domains' (2010a, p. 4). Along with two other criticisms (the overgeneration and absoluteness objections), it led him to an invariantist proposal based on the notion of strong homomorphism rather than isomorphism. We shall consider Feferman's strong-homomorphism invariantism in §8.3.[5]

Should the heterogeneity objection trouble the isomorphism invariantist? In the spirit of giving the objection a sympathetic hearing, let's try and see how we might modify isomorphism invariance to avoid some heterogeneity. We suggest that the way to do so is to take logical relations to be not all and only the isomorphism-invariant ones but the *uniformly*-isomorphism-invariant ones. Informally, a relation is uniformly-isomorphism-invariant iff its interpretation in any submodel is obtained by restricting its interpretation in the original model. Even more

---

[3] Strictly speaking, we only needed one direction, that if a relation is isomorphism-invariant then it is logical.

[4] See §4.6 for the definition of singular and regular cardinals.

[5] Other discussions of heterogeneity include Sher (2016, pp. 311–12), with which we are largely in agreement.

informally: it's the 'same' relation on all domains. The more formal way to express the formal condition, where $\mathcal{M}_1$ is a model and $\mathcal{M}_2$ its submodel (with domain $D(\mathcal{M}_2)$), is: $R^{\mathcal{M}_1}|_{\mathcal{M}_2} = R^{\mathcal{M}_2}$, that is, R's interpretation on $\mathcal{M}_1$ restricted to $\mathcal{M}_2$ is R's interpretation on $\mathcal{M}_2$. Suppose, then, that we lay down the following sufficient condition on what it is to be a logical relation:

Uniform-Isomorphism Invariance Any relation that is uniformly-isomorphism invariant is logical.

A corresponding necessary condition on the one true logic follows naturally:

OTL must contain a logical constant for each uniformly-isomorphism-invariant relation.

In the spirit of sympathetic exploration, let's run with this idea and see what follows from it. As mentioned in Chapter 6, Boolean operations may be incorporated into our extensional framework by considering them as functions from finite tuples of truth-values (T and F) to these values themselves. Generalized Boolean operations—involving κ-ary conjunction and disjunction—may be accommodated in the same way. These operations then drop out as uniformly-isomorphism invariant, since their values do not depend on the model's particular features. For example, $\aleph_0$-conjunction is uniformly-isomorphism-invariant because it expresses the same operation on the countably many truth-values $v_1, \ldots, v_n, \ldots$ in any model $\mathcal{M}_1$ and thus any submodel $\mathcal{M}_2$ of $\mathcal{M}_1$. $\aleph_0$-conjunction is simply the operation which takes $v_1, \ldots, v_n, \ldots$ to T iff all the v_i are T.

Let's turn our attention to the quantifiers. We may take these to denote either unary or binary properties of unary properties of type (i); in other words, we may take the existential and quantifiers to denote relations of type $((i))$ or type $((i),(i))$.[6] Suppose we take them to be unary (we return to the binary option below). As a gentle illustration, consider what happens when we restrict the usual existential quantifier from a three-element domain to a two-element one. Let $\mathcal{M}_1$ be a model with the three-element domain $D(\mathcal{M}_1) = \{a, b, c\}$ and $\mathcal{M}_2$ its submodel with domain $D(\mathcal{M}_2) = \{a, b\}$. The interpretation of the existential quantifier over the first model consists of all its non-empty subsets,

$$\exists^{\mathcal{M}_1} = \{\{a\}, \{b\}, \{c\}, \{a, b\}, \{a, c\}, \{b, c\}, \{a, b, c\}\},$$

and its interpretation over the second model is similarly

$$\exists^{\mathcal{M}_2} = \{\{a\}, \{b\}, \{a, b\}\}.$$

---

[6] See Chapter 4 for the finite-type hierarchy.

$\exists^{\mathcal{M}_1}|_{\mathcal{M}_2}$ is the subset of $\exists^{\mathcal{M}_1}$ whose elements are all the non-empty subsets of $D(\mathcal{M}_2) = \{a, b\}$. It is immediate, then, that $\exists^{\mathcal{M}_1}|_{\mathcal{M}_2} = \exists^{\mathcal{M}_2}$.

The more general argument, for existential quantifiers of different types, is almost as easy. For simplicity and to avoid notational clutter, we give an informal argument for a single existential quantifier. Suppose $\exists$ is of type $((\tau))$, and P is a property of type (τ). (In the example just given, we considered the existential quantifier of type $((i))$ whose arguments are of type (i)—properties of individuals—so that $\tau = i$.) Then $P \in \exists^{\mathcal{M}_1}$ iff P is not the empty set of type (τ) (in $\mathcal{M}_1$), and likewise $Q \in \exists^{\mathcal{M}_2}$ iff Q is not the empty set of type (τ) (in $\mathcal{M}_2$). Thus $\exists^{\mathcal{M}_1}|_{\mathcal{M}_2}$ consists of all and only the non-empty sets of type (τ) over $D(\mathcal{M}_2)$. This is precisely $\exists^{\mathcal{M}_2}$. It is also easy to check that infinitary existential quantification meets the uniform-isomorphism-invariantism condition.

Another easy argument shows that identity is a uniformly-isomorphism-invariant relation. For the extension of the identity relation of type (i, i) on $\mathcal{M}_1$ is the subset of $D(\mathcal{M}_1) \times D(\mathcal{M}_1)$ consisting of all and only ordered pairs of the form $\langle x, x \rangle$, with $x \in D(\mathcal{M}_1)$. Restricted to $\mathcal{M}_2$, this is simply the identity relation's extension on $\mathcal{M}_2$. Similarly for the identity relation of any other type. Conversely, it is easy to check that the regular/singular cardinal quantifier encountered earlier is not uniformly-isomorphism invariant. Uniform-isomorphism invariance is intended to rule out the logicality of this sort of heterogeneous quantifier, and it does the job.

An equally straightforward argument shows that infinitary existential quantifiers and generalized Boolean operations are both uniform-isomorphism invariant. Hence on this approach OTL also has sort $\infty\infty$.[7] For, as just observed, negation and κ-ary disjunction (as well as κ-ary conjunction) are uniformly-isomorphism-invariant operations. Their isomorphism invariance follows from the results stated in Chapter 6, and their uniform-isomorphism invariance by the results just sketched. By a straightforward extension of the argument in the previous-to-last paragraph, κ-ary existential quantification is also uniformly-isomorphism-invariant. On uniform-isomorphism invariance, then, OTL must have a logical constant for negation, κ-ary disjunction and κ-ary existential quantification for any cardinal κ. So if OTL is a sublogic of FTT$_{\infty\infty}$ its sort must be $\infty\infty$, and more generally its sort should be maximally infinitary. In other words, the L∞G∞S Hypothesis follows from uniform-isomorphism-invariance just as it did from isomorphism invariance.

As far as our purposes in this book go, we could stop here. Our goal is to defend the L∞G∞S Hypothesis, and since either form of invariantism—isomorphism invariance or uniform-isomorphism invariance—does the job, we need not choose between them. Still, it is worth looking at uniform-isomorphism-invariance with

---

[7] See Chapter 4 for the definition of sort.

a more critical eye. To make a start on this, we observe that, like logicians of pretty much any stripe, we take closure under definability to be a non-negotiable constraint on a candidate foundational logic. So we lay down the condition:

OTL is closed under definability.

Closure under definability, however, brings with it a good amount of heterogeneity. Consider first pure FOL, that is, FOL with only 'logical' vocabulary, as it's usually put. Pure FOL, which we may take to include the connectives $\wedge, \vee$, and $\neg$ as well as $\forall, \exists, =$ (and variables and parentheses), can define a whole host of Boolean connectives. It can for instance define $p * q$ by

$$p * q =_{def} (\exists_{=2} \to p \vee q) \wedge (\neg\exists_{=2} \to p \wedge q).$$

Here $\exists_{=2}$ is a pure FOL-sentence true in all and only domains of size 2. The binary operator $*$ is thus defined as disjunction in two-membered domains and conjunction in all other ones. Although heterogeneous, $*$ is definable from connectives that are *uniform*, that is, non-heterogeneous.

How much heterogeneity does FOL exhibit? To investigate this, let's restrict attention to *pure* FOL. Call a *finite-size-sentence* any pure FOL-sentence formalizing 'There are at least n things', 'There are at most n things', or 'There are exactly n things' for finite n. An *identity formula* is of the form $x = y$ for first-order variables x and y. A standard inductive argument shows that any pure FOL-formula is FOL-equivalent to a Boolean combination of finite-size-sentences and identity formulas.[8] This implies that any pure FOL-sentence is FOL-equivalent to a Boolean combination of finite-size-sentences. Suppose then that ϕ is a pure FOL-sentence, and that C_ϕ is the class of cardinals in which ϕ is true.[9] Then by the result just stated, C_ϕ is either a finite class or a cofinite class, where a cofinite class is one whose complement is finite.[10]

Pure FOL thus admits a fair amount of heterogeneity. The same applies to pure $\mathcal{L}_{\infty\infty}$. The simple result just outlined for pure FOL finds an analogue in $\mathcal{L}_{\infty\infty}$. Call a *size sentence* any pure $\mathcal{L}_{\infty\infty}$-sentence formalizing 'There are at least κ things',

---

[8] Ch. 1 of Chang and Keisler (1990).

[9] It is easy to see that ϕ is true in a model with domain of a particular cardinality iff it is true in all models of this cardinality.

[10] Something similar can be said for impure FOL. The upward Löwenheim-Skolem theorem tells us that any FOL-sentence with an infinite model has a model of all infinite cardinalities. Hence there can be no FOL-sentence $\sigma(p_1, \ldots, p_n)$ that defines an n-place Boolean operator that acts one way, as O_P say, on all models of size meeting some property P and another way, O_{NP} say, on all models of size that lack property P, *where the class of P-cardinals and the class of NP-cardinals are both unbounded*. (And where O_P and O_{NP} disagree on at least one n-tuple of truth-values.) For since O_P is by assumption FOL-definable, the sentence $\sigma(p_1, \ldots, p_n) \leftrightarrow O_P(p_1, \ldots, p_n)$ would be true in all and only models with property P, contradicting the upward Löwenheim-Skolem Theorem.

'There are at most κ things', or 'There are exactly κ things' for any cardinal κ (not just finite ones); similarly, identity formulas are of the form $x = y$ and generalized Boolean combinations allow applications of negation as well as κ-ary conjunction or disjunction, for any κ. Then any pure $\mathcal{L}_{\infty\infty}$ sentence is $\mathcal{L}_{\infty\infty}$-equivalent to a generalized Boolean combination of size sentences.[11] Now a straightforward argument shows that, for any generalized Boolean combination σ of size sentences, there is a cardinal κ such that *either* all of σ's models are of size no more than κ, *or* any model of size greater than κ is a model of σ. In this precise sense, then, pure $\mathcal{L}_{\infty\infty}$ also admits a fair amount of heterogeneity.

Returning to the philosophical thread, we have seen that the logical constants should be closed under definability. This means that there must be some very heterogeneous and *ad hoc*-looking logical constants. An example is $\forall_{=17}/\exists_{\neq17}$, which behaves like the universal quantifier on domains of cardinality 17 and like the existential quantifier on domains of any other cardinality. This strongly suggests that any intuition to the effect that $\forall_{reg}/\exists_{sing}$ is too heterogeneous to be a logical relation is based on a misconception and shouldn't be pandered to. In fact, given that the $\forall_{=17}/\exists_{\neq17}$ quantifier has to be a logical constant for anyone who accepts that at least first-order logic is logic, it seems *ad hoc not* to count $\forall_{reg}/\exists_{sing}$ as a logical constant too. To turn the heterogeneity objection on its head: classifying $\forall_{reg}/\exists_{sing}$ as a logical constant should be considered a desideratum for an account of logical constanthood.[12]

Two more points can be made against the uniform-isomorphism version of invariantism. The first is that, as we have seen, the class of uniform-isomorphism-invariant expressions (meaning expressions that denote a uniform-isomorphism-invariant relation) is not closed under definability. This, it may be said, defeasibly tells against uniform-isomorphism invariantism's claim to be closely connected to logicality, be it the logicality of extensions or of constants. Of course, it may still be a sufficient condition for logicality, but it would be preferable to derive it from a more fundamental characterizing condition. In contrast, isomorphism invariance does offer a unified necessary and sufficient account of logical relations.[13]

Second, the unary universal quantifier is not uniform-isomorphism-invariant. Consider the unary universal quantifier of type $((i))$, which is a unary property of unary properties. (The discussion generalizes to unary universal quantifiers of other types.) When $D(\mathcal{M}_1) = \{a, b, c\}$ and $D(\mathcal{M}_2) = \{a, b\}$, as above, then $\forall^{\mathcal{M}_1} = \{\{a, b, c\}\}$, and the restriction of this set to $\mathcal{M}_2$ is the empty set, since $\{a, b, c\}$ is not a subset of $D(\mathcal{M}_2)$. Yet $\forall^{\mathcal{M}_2} = \{\{a, b\}\}$. Now uniform-isomorphism invariantism was introduced as a sufficient condition for logicality, not a necessary one. And clearly, the universal quantifier may be defined by means of the existential quantifier and

---

[11] The argument is an analogue of the FOL-argument.
[12] In agreement with Sher (2016, p. 312).
[13] Although note that we only needed one direction of it in Part II.

negation, both of which are uniform-isomorphism-invariant expressions. Still, the fact that the universal quantifier, a paradigm of a logical constant, fails the uniform-isomorphism invariance test speaks ill of the criterion. It casts doubt on the link between uniform-isomorphism invariance and logicality.[14]

In summary, in this section we introduced the heterogeneity objection to isomorphism invariance, advanced by Feferman and others in his wake. To avoid it, we considered spelling out invariantism in terms of uniform-isomorphism invariance rather than isomorphism invariance. We saw that this approach avoids only a little heterogeneity and is unattractive in other ways. So it may not be a very promising form of invariantism. Of course, as far the defence of the L∞G∞S Hypothesis goes, either of these forms of invariantism will do; we need not choose between them. We move on to two stricter invariantisms, which have much more drastic consequences for our main thesis.

8.3 Feferman's invariantism

In a series of publications, Solomon Feferman put forward an invariantist account of logical constants as a rival to isomorphism invariance.[15] Restricting attention to notions of type level 2 (that is of type $((i))$ or $((i),(i))$ and so on), Feferman proposed a *strong-homomorphism-invariance* criterion of logicality. This is what we now present.[16]

Suppose the operation Q we are interested in is of type level 2 and arity n, so that it takes arguments $R_1, \ldots, R_n$, that is, n relations over some domain D. To say that Q is of type 2 is to say that each R_j is of type (i) or (i,i) or (i,i,i) and so on, where i is the type of individuals; we use the letter 'Q' as such operations may be thought of as generalized quantifiers. Where $P_1, \ldots, P_n$ are n predicate symbols of the appropriate adicity, we can think of a model as consisting of a domain D and interpretations $R_1, \ldots, R_n$ of the n predicate symbols $P_1, \ldots, P_n$, where R_i's arity matches P_i's adicity. In short, a model is an $n+1$ tuple $\langle D, R_1, \ldots, R_n \rangle$. The formula

---

[14] We note a caveat, though. If we take (the relation denoted by) ∀ as a binary rather than a unary property of unary properties, the situation changes. Indeed, in linguistics, it is customary to regard the unary universal quantifier as a restriction of the binary one, and more generally to regard quantifiers of type $((i),(i))$ as more basic than those of type $((i))$ (see e.g. Peters and Westerståhl (2006) *passim*; the point is explicitly made on p. 12). The truth-conditions of 'All Fs are Gs' are given, naturally, by specifying that the extension of F complement G's extension (i.e. $F \setminus G$) is empty. So given a model $\mathcal{M}_1$, $\forall^{\mathcal{M}_1}$ consists of all ordered pairs $\langle P_1, P_2 \rangle$, where $P_1, P_2 \subseteq D(\mathcal{M}_1)$, such that $P_1 \subseteq P_2$. This construal of the universal quantifier manifestly meets the uniformity condition, since $\forall^{\mathcal{M}_1}|_{\mathcal{M}_2}$ consists of all the ordered pairs $\langle P_1, P_2 \rangle$, with $P_1, P_2 \subseteq D(\mathcal{M}_2)$, satisfying the same condition, i.e. $P_1 \subseteq P_2$. The analogous truth-condition for the existential quantifier thus specified—capturing the form 'Some Fs are Gs'—should be $P_1 \cap P_2 \neq \emptyset$. This is also easily seen to be uniform. Similarly for 'Most Fs are Gs', whose extension consists of the ordered pairs $\langle P_1, P_2 \rangle$ with $P_1 \cap P_2 > P_1 \setminus P_2$.

[15] See especially Feferman (2010a).

[16] Sher (2013, p. 196 fn. 50) reports that in 2011, five years prior to his death, Feferman had given up this alternative form of invariantism but still stood by his criticisms of isomorphism invariance.

$\phi(P_1, \ldots, P_n)$ is said to define Q_D, Q's interpretation over D, just when, for any $R_1, \ldots, R_n$ of the appropriate arity, $Q_D(R_1, \ldots, R_n)$ iff ϕ is true in the model whose domain is D and which interprets $P_1, \ldots, P_n$ as $R_1, \ldots, R_n$.

Now on to the crucial notion. A strong homomorphism h between models $\langle D, R_1, \ldots R_n, \rangle$ and $\langle D^*, R_1^*, \ldots, R_n^* \rangle$ is a map from D to D^* that, in Feferman's application, is stipulated to be *onto* and such that, for each i from 1 to n and any sequence $x_1, \ldots x_k$ of individuals drawn from D: $R_i(x_1, \ldots, x_k)$ iff $R_i^*(h(x_1), \ldots, h(x_k))$.[17] Given these definitions, the operation Q is strong-homomorphism-invariant just when, for any such homomorphism h, Q holds of the sequence $R_1, \ldots R_n$ in the first model (with domain D) iff it holds of the sequence $R_1^*, \ldots R_n^*$ in the second model (with domain D^*).[18]

It is an immediate consequence of Feferman's criterion that most finite cardinality quantifiers are not logical, as Feferman himself noted. As an illustration, let the domain D be $\{a, b\}$, where $a \neq b$, and the domain D^* be $\{c\}$. There is of course only one strong homomorphism from D to D^*, which maps each of a and b to c. Now suppose we consider the models $\langle D, R \rangle$ and $\langle D^*, R^* \rangle$ in which R and R^* are unary properties that hold of all the entities in their respective domains, that is, $R = D$ and $R^* = D^*$. Let $\exists_{\geq 2}$ be the property of type $((i))$ which holds of a property of type (i) iff the latter holds of at least two individuals in the domain. It's easy to see that $\exists_{\geq 2}(R)$ (in the first model) but that $\exists_{\geq 2}(R^*)$ does not hold in the second model. A similar argument shows that $\exists_\kappa$ is not strong-homomorphism-invariant, where κ is any cardinal ≥ 2 and $\exists_\kappa(R)$ holds iff R holds of at least κ individuals in the domain.

Equally noteworthy is the fact that identity is not strong-homomorphism invariant. Since identity is a relation of type (i, i) and is not of type level 2, the criterion of logicality just given does not apply to it. But we can design a strong-homomorphism-invariance criterion for entities of identity's type in the spirit of Feferman's original one. We may stipulate, for example, that $\mathcal{M}$ is strong-homomorphism invariant to $\mathcal{M}^*$ just when there is a map from $\mathcal{M}$'s domain to $\mathcal{M}^*$'s that is onto and respects the interpretation of any constant and function symbols. Then it is easy to see, using the previous example, that the interpretation of identity under the homomorphism $h : \{a, b\} \to \{c\}$ is not preserved: it is false that all pairs $\langle x, y \rangle$, with $x, y \in D$, stand in the identity relation in the first model iff $\langle h(x), h(y) \rangle$ stand in the identity relation in the second model, since $a \neq b$ but $h(a) = h(b)$. Thus identity is not logical on the strong-homomorphism-invariance account. Feferman is entirely clear about this consequence and accepts it.

---

[17] Here k is R_i's arity. The obvious clauses for function symbols and constants may be added.

[18] We are skating over some exegetical matters irrelevant to our philosophical overview. In particular, Feferman first proposed strong-homomorphism invariance as a criterion for operations of unary type, i.e. of type $((i))$, and considered their closure under λ-definability. It seems to us, as it did to Feferman later, that there is no good reason to restrict invariantism to operations of this particular type and no other. See Bonnay (2008, pp. 43–4) for detailed criticism of Feferman's earlier proposal.

The strong-homomorphism-invariance criterion of logicality seems to us much more problematic than one based on isomorphism invariance. An obvious and immediate reason is that it rules out identity as logical. This runs counter to mathematical and logical practice, in which the status of identity as logical is well-entrenched. It also runs counter to the invariantist's motivation. We and most others believe that logic should be insensitive to the particular nature of objects though sensitive to their existence or obliteration—in other words, to their identity. Like (virtually) everyone else, we take first-order logic with identity to be a part of logic, and see any proposal that would curtail the one true logic to a first-order fragment as evidently misguided. Moreover, in second-order logic, first-order identity is definable,[19] so that anyone who takes second-order quantifiers to be logical has an even stronger reason for classifying identity as logical. Since Feferman himself does not really offer any arguments against taking identity to be logical, we cannot see any grounds for overturning this conviction. If even identity's logicality is up for grabs, we find it hard to see how *any* headway can be made on the question of the logical constants.

A closely related point is that finite cardinality quantifiers (e.g. of the form 'there exist at least n' for finite n) should count as logical. All these cardinality quantifiers are definable in first-order logic, which is why they are usually regarded as logical. Yet, as we saw, quantifiers such as 'there exist at least n' for $n \geq 2$ are not deemed logical by Feferman's account.

Naturally, this second point is highly dependent on the previous one, since first-order-definable cardinality quantifiers other than the existential are definable from it by essential use of identity. But the point is nonetheless separate, since the reasons for taking finite cardinality quantifiers to be logical do not all flow from the logicality of identity. There is, for example, an independent degree of plausibility to taking 'there are at least two' to be logical if one takes 'there is at least one' to be logical. Inferentially speaking, one can give introduction and elimination rules for finite cardinality quantifiers, without relying on identity.[20] Granted identity, the rules are just like the existential ones, bar this complexity.

We note in passing an *ad hominem* point against Feferman, who ends one of his last papers on this subject with the following observation:

And that returns us to the traditional conception of logic as the study of the *forms of correct reasoning, of what invariably leads from truths to truths*. Despite

---

[19] Define $t_1 = t_2$, where t_1 and t_2 are terms by $\forall X(Xt_1 \leftrightarrow Xt_2)$. This can be done at every type.

[20] For example, 'there are at least two things' may be inferred from '$Fa \wedge \neg Fb$'. In a sufficiently rich language, this latter sentence may be inferred from 'there are at least two things' for some predicate F and constants a and b. Or even if the language is not sufficiently rich, it may be inferred from 'there are at least two things' if we are careful not to assume any further facts about the predicate letter F beyond those warranted by this premise (compare the introduction of constants from existential claims in natural-deduction systems).

the various appealing results above, and despite my personal feeling that the logical operations do not go beyond those represented in FOL, I do not find the various arguments for logicality based on any of the invariance notions considered here convincing in their own right. In my view, the semantical and syntactic (inferential-theoretic) approaches are complementary to each other, and a proper explanation of what are logical notions and of what is logic—if there is to be one—will have to take both into account.

(Feferman 2010a, p. 17; italics in the original).

As is clear from this passage, Feferman takes it to be a virtue of strong-homomorphism invariance, as opposed to isomorphism invariance, that it leads to a *rapprochement* with the proof-theoretic (inferentialist) approach to the logical constants. Extensionally speaking, the set of constants the former vindicates is much more similar to the constants inferentialists accept than those under isomorphism invariance. This is presumably what also underlies his interest in logics which satisfy the condition integral to Lindström's Second Theorem: that the set of logical validities be recursively enumerable (Feferman 2010a, pp. 14, 18). From a model-theoretic (semanticist) perspective, however, the motivation for this condition is missing. Yet from a perspective which takes proof theory seriously, at least as currently practised and understood, both identity and finite cardinality quantifiers are logical. And as we saw in §5.2, if you are moved by inferentialist considerations, or at least take the confluence of semantic and syntactic characterizations as a sign of logicality, then you have reason to rule in 'there exist uncountably many' as logical. Our own approach cuts loose from inferentialist moorings. But if one seeks a 'complementary' approach that brings together semanticists and inferentialists, it is hard to see how strong-homomorphism invariance can be the right account of logicality.

A third problem for strong-homomorphism invariance is its motivation. A common and deeply engrained idea is that logic should be insensitive to the identity of objects. This is the idea on which Tarski based his 1986 account of logical constants. It is an idea which surfaces again and again in the history of the philosophy of logic, and is supported by logical practice. A thought that is much less intuitive and much less common is this: in logic, not only does it not matter what type of objects we are reasoning about, it also does not matter how many of them there are. In reasoning logically about, say, three llamas, not only could we imagine them to be horses without a difference in logical content (assuming we keep track of the change from llamas to horses throughout), but we could equally imagine them to be a single llama. Feferman's account is committed to the latter idea, which goes squarely against our conception of logicality.

Even if that last thought does not strike you as counterintuitive, a problem of implementation plagues Feferman's account. The issue has do with the adjective *strong* in 'strong-isomorphism invariance', marking the dependence on *onto* homomorphisms. *Why* should homomorphisms be onto? If the idea is to liberalize isomorphism invariance so that logic is insensitive to the number of objects in the domain, why shouldn't we consider general homomorphisms, which may or may not be onto? Or to put it in more homely terms, consider a domain that includes three llamas as well as some finite number of horses. We are supposed to imagine that, as far as a logical characterization of the domain is concerned, the llamas might as well be two or one, without changing the amount of horses. But why could there not be four or five llamas, without there being any fewer horses? Why should the domains of comparison over which logical operations must be constant always be of *no greater* size than the initial domain? We cannot see any good reason— other than the wish to not further restrict the scope of the logical—to stop at Feferman's stopping point. And notice that if homomorphisms need not be onto then the universal quantifier will *not* be logical. Construed as the second-order property $Q_\forall$, $Q_\forall R$ obtains in D when $R = D$; yet if $h : D \to D*$ is a non-surjective homomorphism and for $x \in D^*$, $R^*(x)$ iff $x \in h(D)$, then $Q_\forall R^*$ does not obtain in D^* (since h is not surjective, $h(R) = R^* \neq D^*$).

Finally, we note in passing an issue to do with empty domains which is usually fudged in contemporary logic. Usually, in logic, we take domains to be non-empty, for the reason that this yields a technically simpler account, as is familiar. Yet many logicians would maintain that this is *only* for technical convenience. The correct, model-theoretic, account of logical consequence should in good philosophical conscience allow for empty domains. Accordingly, a statement such as 'there is at least one thing', which comes out as a logical truth if models are non-empty, should not really be considered a logical truth.[21]

It is worth noting that the isomorphism-invariance approach can quite happily accommodate empty domains. To see this, note first that all operations are isomorphism-invariant on the empty domain or what comes to the same thing the empty model, whose domain is empty and all of whose relations, of any type, are empty. Trivially so, since there is only one empty model, so all operations are isomorphism invariant over models of this cardinality. In other words, allowing empty models does not affect which relations are isomorphism invariant. Iso-morphism invariantists can fully take on board the apparently reasonable point that models may be empty: it does not alter their account of logical relations or constants. The same applies to uniform-isomorphism invariance, since the

---

[21] The technical challenge is then to implement this idea. The sorts of logics one could appeal to here are Quine's (1954) *inclusive* logic or Meyer and Lambert's (1968) *universally free* logic, both of which countenance the empty domain and so rule out the logical truth of e.g. $\exists x(x = x)$. Oliver and Smiley (2013, §11.1) show that there are three main reasons given in the literature to justify rejection of the empty domain: first, it is an anomalous case that can be ignored; second, it is a freak case that should be avoided; third, it introduces serious technical difficulties. Oliver and Smiley argue that all of these reasons are mistaken. We will not rehearse their arguments here but recommend that the reader suspicious of empty domains should see Oliver and Smiley on the subject.

restriction of any relation on a model $\mathcal{M}$ to its empty submodel is empty and therefore identical to its interpretation on the empty submodel.

Is the same true of strong-homomorphism invariance? It is a simple mathematical fact that there are no functions from a non-empty set to the empty set. (A function from X to Y assigns to each element x in X an element y of Y; if X is non-empty and Y is empty there are no such functions.) Since a homomorphism from a domain D to a domain D^* is just a special kind of function (total mapping), it is immediate that there are no homomorphisms from a non-empty domain to an empty one. There is thus a clear sense in which Feferman's account ignores empty models, precisely because there are no strong homomorphisms with the empty domain as their codomain. Although the account is not inconsistent with the existence of empty models, they play no role in the account: they are not included in the range of models that determine logicality because there are no strong homomorphisms to them. So there is a clear sense in which strong-homomorphism invariantists do not take empty models seriously.

In sum, we have presented several objections for Feferman's invariantism. The first is that it counts identity as non-logical. The second is that the finite cardinality quantifiers come out as non-logical. The third is that its motivation is flawed: basing invariantism on *strong* homomorphisms is highly contestable. And if we liberalize the criterion by dropping the onto condition, another undue consequence follows. We also noted a potential fourth problem, to do with empty domains. As we see it, these are all strong reasons for rejecting strong-homomorphism invariance.

8.4　Bonnay's invariantism

We turn more briefly to another invariantist account put forward by Denis Bonnay (2008). Despite the paper's other qualities, we believe the main invariantist proposal it puts forward is flawed, patently so. We *summarily* describe Bonnay's proposed invariantism (potential-isomorphism invariance), his motivation for it, and some reasons to reject it.

Starting with Bonnay's motivating argument, the first of these is strikingly less intuitive and compelling than the corresponding argument behind isomorphism invariance. As Bonnay himself remarks, 'Our revised arguments have lost some of the elegant simplicity of the formality and the generality arguments [for isomorphism invariance]. We think that they could be rephrased in more vernacular terms, but at the cost of either clarity or brevity' (2008, p. 69). Sher (2013, p. 196) couches the same criticism of both Feferman and Bonnay in diplomatic terms: '... not every criterion of logicality is equally conducive to a unified, substantive, theoretical foundation for logic'. The arguments for isomorphism invariance are much clearer and briefer than the convoluted ones for potential-isomorphism invariance, which points to something important about the notions they each promote.

As an illustration, take the first of Bonnay's two main arguments for his potential-isomorphism invariantism, the *Mild Generality* argument.[22] The argument is highly disjunctive and takes it as an unexplained primitive fact that the truth-functions, functional application, and first-order existential quantification are logical operations. It is thus a curious hybrid of two different sorts of motivations: an invariantist one, which seeks to explicate logical notions as invariant over similar types of models; and a first-orderist one which takes the logical operations to be the closure under definability of the usual operations taken as basic in first-order logic.

Bonnay does have a point that a generality argument for isomorphism invariance cannot be unrestricted. Unrestricted generality would correspond to invariance under any similarity relation, going far beyond isomorphism invariance. As we saw in §8.3, Feferman takes strong-homomorphisms as the basis of his account, and one could liberalize even this and consider properties invariant not under isomorphisms but under homomorphisms more generally. This is the thought that lies behind Bonnay's *Mild Generality* argument: applied to the hilt, generality would imply that virtually nothing is logical.

We, however, believe that there is a simplicity and unity to the isomorphism-invariance account of logical extensions (which Bonnay recognizes) that recommends it. It is by far the simplest, most unified and theoretically powerful explication of logicality. To see this, compare Bonnay's *Mild Generality* argument, or the absence of any principled invariantist criterion for taking FOL or SOL to be the one true logic. By normal abductive criteria, isomorphism invariance is much the best explication. Moreover, isomorphism invariance fits hand in glove with our principal motivation for it, namely topic-neutrality. As we saw in §8.3, topic-neutral vocabulary ignores the particular natures of the things under discussion; but it does not ignore their number. So we are not swayed by the criticism of isomorphism invariance that animates Bonnay's first argument: just because can you go beyond a position does not mean you should—it may well be, as in this instance, not merely an excellent stopping point but the best one.

---

[22] Which runs as follows (Bonnay 2008, p. 59):

MG1 Logic deals with only very general notions, but not only with trivial notions.

MG2 The truth-functions, functional application, and first-order existential quantification are logical operators.

MG3 The good notion of invariance for logicality is to be provided by a similarity relation S such that S is closed under definability.

MG4 The good notion of invariance for logicality is to be provided by the lowest [weakest] similarity relation compatible with MG2 and MG3.

So: The logical notions are the Iso_p-invariant [potential-isomorphism-invariant] notions.

Bonnay (2008, p. 53) proves the argument's conclusion from its premises.

The second reason is that Bonnay's invariantism is motivated by the second and fourth objections to isomorphism invariantism: the overgeneration objection and the absoluteness objection. We shall examine these in Chapters 9 and 10 respectively. Bonnay himself is explicit that the point of his approach is to find a form of invariantism that 'does not overgenerate as badly' as isomorphism invariance (2008, p. 39). Indeed, a version of this point is the first premise in the second of his two main arguments for his potential-isomorphism invariantism.[23] Its third premise is simply a version of the point underlying the absoluteness objection. As we said, we'll deal with both these objections in a broader context in later chapters.

Turning to Bonnay's conclusion, then, what is potential-isomorphism invariantism? A potential isomorphism I between two structures $\mathcal{M}$ and $\mathcal{M}'$ is a non-empty set of partial isomorphisms such that for all $f \in I$ and a in the domain of $\mathcal{M}$ (respectively b in the domain of $\mathcal{M}'$) there is a $g \in I$ with $f \subseteq g$ and $a \in dom(g)$ (respectively $b \in ran(g)$). A partial isomorphism $f \in I$ between $\mathcal{M}$ and $\mathcal{M}'$ is an isomorphism between a substructure of $\mathcal{M}$ and a substructure of $\mathcal{M}'$. The resulting invariantist account is potential-isomorphism invariantism.

We note first that Bonnay's invariantism is based on a much more complicated notion of similarity between structures than isomorphism. What drives invariantism, and the semanticist approach to the logical constants more generally, is the idea that logical operations act the same way on similar domains. It is very natural to understand 'similar' more precisely as 'isomorphic', and much less natural to understand it as 'potentially isomorphic'. The latter notion is more sophisticated and complicated than that of isomorphism, and indeed deploys the notion of an isomorphism as part of its definition.

Second, we note that the potential-isomorphism account patently fails to achieve one of the aims driving it, namely to avoid 'mathematical content'. To take just one example, the account makes the notion 'structure potentially isomorphic to the real numbers with the less-than relation' logical. This class includes not

---

[23] That is, the *Lack of Content* argument (Bonnay 2008, p. 60). This argument is based on an order $\leq$ on similarity relations between models, so that $S \leq S^*$ iff for any two models $\mathcal{M}_1$ and $\mathcal{M}_2$, if $\mathcal{M}_1$ and $\mathcal{M}_2$ are S^*-related then they are S-related. Intuitively, S^* draws finer distinctions among models than S does. *Iso* is the similarity relation of isomorphism invariance and Iso_p that of potential-isomorphism invariance. The second argument now runs as follows:

LC1 Logic deals with notions which are deprived of nonformal content and of problematic set-theoretic content.

LC2 The good notion of invariance for logicality is to be provided by a similarity relation S such that $S \leq Iso$.

LC3 The good notion of invariance for logicality is to be provided by a similarity relation S such that S is absolute with respect to ZFC.

LC4 The good notion of invariance for logicality is to be provided by the greatest similarity relation S satisfying LC2 and LC3.

So: The logical notions are the Iso_p-invariant notions.

A theorem of Barwise's shows that the conclusion follows from the premises (Bonnay 2008, p. 57).

just structures isomorphic to the real numbers under less-than but also structures such as the rational numbers under less than.

A third problem with the view is that it draws unacceptable distinctions. For example, it takes the quantifier 'there exist at least $\aleph_0$-many' to be logical but not the quantifier 'there exist exactly $\aleph_0$-many', as Bonnay himself points out (2008, p. 61). Now, first-orderists, who take FOL to be the one true logic, think there is a gulf between finite and infinite, so that finite cardinality quantifiers are logical but infinitary ones are not.[24] Although we are opposed to first-orderism, we recognize that this position has a degree of stability and coherence to it. Not so with the idea that some infinitary quantifiers are logical whereas other infinitary quantifiers are not. To be infinite is not to be bijectible with any proper initial segment of the natural numbers; to be countably infinite is to be bijectible with the natural numbers. We do not see here any gulf on which to found a distinction between logicality and non-logicality. Nothing in logic, mathematics or philosophy justifies it.

Fourth, it turns out by a theorem of Barwise's that an operator Q is potential-isomorphism-invariant iff for any structure M, Q's action on the structure is definable in $\mathcal{L}_{\infty\omega}$.[25] Now $\mathcal{L}_{\infty\omega}$ is as infinitary as can be in *one* respect, in that it allows arbitrary conjunctions and disjunctions and properties; but it is also finitary in *another*, in that it allows quantification over only finitely many variables. In contrast, $\mathcal{L}_{\infty\infty}$ is consistently infinitary, and $\mathcal{L}_{\omega\omega}$ is consistently finitary. Why allow infinitary Boolean combinations of formulas, but not infinitary quantification? On the potential-isomorphism account, the resulting logic is a Frankensteinish hybrid of the infinitary-through-and-through $\mathcal{L}_{\infty\infty}$ and the finitary-through-and-through $\mathcal{L}_{\omega\omega}$.

8.5 Conclusion

We introduced the first of the four standard objections to isomorphism invariance in this chapter: the heterogeneity objection (as we labelled it). Prompted by this objection, we developed a new form of invariantism, uniform-isomorphism invariance. But we saw in §8.2 that heterogeneity is an inevitable by-product of closure under definability. On that basis, we argued that an invariantist should embrace, rather than reject, heterogeneity. Although we are not tempted by uniform-isomorphism invariance, whether we plumb for it or the more traditional isomorphism invariance does not in the end matter for our purposes. Both these invariantisms vindicate our thesis that the one true logic is maximally infinitary.

---

[24] The former, but not the latter, are FOL-definable.
[25] Bonnay (2008, p. 62). Recall the definitions of infinitary logics in Chapter 4.

We then looked at two rather different invariantisms: Feferman's strong-homomorphism invariantism and Bonnay's potential-isomorphism invariantism. We argued that Feferman's is not as well-motivated as the isomorphism-invariance account we relied on in Part II, and mentioned more briefly that the same applies to Bonnay's. We also saw that both Feferman's and Bonnay's invariantisms have unacceptable consequences. We now move on to the second objection to isomorphism invariance: overgeneration. As this is the objection critics of isomorphism invariance take to be the most fundamental, we shall examine it in considerable detail.

9
The Overgeneration Objection

9.1 Introduction

We come now to the second of the four main objections levelled at isomorphism invariance, in many ways the most important and potentially damaging one. The *overgeneration* objection aims to show that the isomorphism-invariance criterion overgenerates by deeming as logical some sub-sentential expressions that are not, and consequently some truths as logical that are not. In particular, isomorphism invariance takes as logical all cardinality quantifiers of the form 'there exists at least κ' or 'there exists at most κ' or 'there exists exactly κ', for any non-zero cardinal κ. The reason these quantifiers count as logical was set out in Chapter 6 and is intuitively obvious: since isomorphisms preserve the cardinality of domains, cardinality quantifiers are all isomorphism invariant and hence, on the criterion, logical.

Why is it problematic to judge these quantifiers as logical? The following quote from Bonnay is representative of the worries some authors have:

> this suggests that Tarski's criterion overgenerates and counts too many operations as logical. First, since the aim is to distinguish the realm of logic proper, a proposal which conflates logical notions and mathematical notions does not seem to be on the right track. (Bonnay 2008, p. 37)

The thought is that quantifiers of these sorts are not logical because to call them so conflates logic and mathematics. On the next page, Bonnay claims that the situation is even worse for isomorphism invariance, since it 'yields a collapse of logic into mathematics' (Bonnay 2008, p. 38). Similarly, Feferman writes that isomorphism invariance 'assimilates logic to mathematics, more specifically to set theory' (Feferman 2010, p. 3). The theme in these claims is that isomorphism invariance overgenerates because it somehow involves a confusion of logical and mathematical notions. In particular, it renders certain mathematical notions logical in a way that is problematic.

When we look for cases of overgeneration in these authors' works, the same example always comes up. Feferman writes that, if we accept isomorphism invariance as our criterion of logical nature, 'we can express the Continuum Hypothesis and many other substantial mathematical propositions as logically determinate sentences' (Feferman 2010a, p. 12). Similarly, Bonnay writes that, on the criterion

One True Logic: A Monist Manifesto. Owen Griffiths and A.C. Paseau, Oxford University Press.
© Owen Griffiths and A.C. Paseau 2022. DOI: 10.1093/oso/9780198829713.003.0009

the quantifier $Q_{\aleph_1}$, which tests whether there are exactly $\aleph_1$ objects satisfying the formula, is logical. Intuitively, something has gone wrong. Being of size $\aleph_1$ is a notion that belongs to set theory, not to logic. . . . it is possible to express in $\mathcal{L}_{\infty,\infty}$ the Continuum Hypothesis and other substantial set-theoretic claims.

(Bonnay 2008, p. 36)

Prior to Feferman and Bonnay, other philosophers had made the same point about second-order logic using the same example of the Continuum Hypothesis (CH). The overgeneration objection to SOL is set out and vigorously defended in Etchemendy (1990). Following Etchemendy, Patricia Blanchette discusses the overgeneration argument in a handbook entry and concludes that it is successful, taking it 'to provide a reason for deeming the usual second-order model-theoretic consequence relation unreliable as an indicator of logical consequence' (2001, p. 129).[1] McGee (1992, p. 273 fn.1) describes Etchemendy's book as providing penetrating criticisms of the model-theoretic conception of logical consequence and truth, to which McGee attempts to add more of his own. A more critical yet still concessive attitude is taken by Shapiro (1998, p. 146), who on the basis of the overgeneration objection to SOL embraces the conclusion that there is no boundary between mathematics and logic. Hanson (1997, p. 397) likewise accepts that we may have 'a sentence with the content of the continuum hypothesis as a logical truth'. Priest (1995, p. 289) concurs with Etchemendy that there is a formula of pure second-order logic which 'expresses' CH, and tries to block the argument by modifying the language of second-order logic. In his own 'rethinking' of the overgeneration objection to SOL (Etchemendy 2008), its originator claims that the argument, along with the other significant points made in his 1990 book, is 'essentially correct' (2008, p. 263). In short, many critics of SOL and isomorphism invariance as a criterion of logicality alike are agreed on this point: commitment to either leads to an 'overgeneration' of logical truths.

Although it is well-known, it is important to be clear about what the Continuum Hypothesis states in order to assess this objection. CH is the interpreted first-order sentence abbreviated to:

CH $2^{\aleph_0} = \aleph_1$; in words: the cardinality of the power set of the first infinite cardinal is the first uncountable cardinal.[2]

CH is a controversial, ZFC-undecidable claim.[3] And now for the allegedly problematic link between set theory, on the one hand, and either SOL or infinitary logics

---

[1] Blanchette (2000, pp. 63–5) also discusses what she calls 'problematic model-theoretic truths' of this ilk.

[2] Strictly speaking, $2^{\aleph_0}$ is the cardinal equinumerous with the set of functions from $\aleph_0$ to $2 = \{0,1\}$. Clearly, $|\mathbb{P}(\aleph_0)| = 2^{\aleph_0}$.

[3] Thinking of ZFC as a formal theory, we could more strictly write that the first-order sentence in the language of set theory of which CH is an interpretation is ZFC-undecidable. This reading may be applied to all such claims below.

such as $\mathcal{L}_{\infty\infty}$, on the other: there are sentences of these logics which are logically true iff CH is true, and other sentences of these same logics which are logically true iff CH is false.

Starting with SOL, let $N(X)$ be a second-order formula that is satisfied just if X is equinumerous with the natural numbers, and $R(Y)$ a second-order formula that is satisfied just if Y is equinumerous with the real numbers. These expressions can all be defined in purely second-order vocabulary.[4] Let $X < Y$ and $X \leq Y$ abbreviate the usual second-order renderings of 'X has cardinality smaller than Y' and 'X has cardinality smaller than or equal to Y', respectively. Using these abbreviations to keep it manageable, consider the (uninterpreted) formula

$$\forall X \forall Y \forall Z ((N(X) \land R(Y) \land X < Z) \to Y \leq Z).$$

Now interpret this formula over a domain of entities, say one that contains at least continuum-many elements, or perhaps the domain of all things. The resulting statement, which we call S, is a second-order logical truth just when CH is true. (More on what S states in §9.2.)

Assuming that CH is true, anyone who takes second-order logic as logic is, therefore, committed to the logical truth of S. Of course, the truth of CH is controversial, but an exactly analogous argument can be run on the assumption that CH is false with another sentence S' that is logically true just if CH is false.

Interpreted sentences of $\mathcal{L}_{\infty\infty}$ that are respectively logically true iff CH is true and logically true iff CH is false may similarly be found. Very briefly, the reason is that $\mathcal{L}_{\infty\infty}$ can define the quantifier $Q_{\aleph_\alpha}$, interpretable as 'there are exactly $\aleph_\alpha$' (for α an ordinal).[5] It can thus express the sentence $Q_{2^{\aleph_0}} x(x = x) \leftrightarrow Q_{\aleph_1} x(x = x)$, which is an $\mathcal{L}_{\infty\infty}$-logical truth iff CH is true. A fortiori, any sublogic of $\mathrm{FTT}_{\infty\infty}$ of sort $\infty\infty$ can define these too,[6] and of course the same is true mutatis mutandis if CH is false. For brevity, we shall focus on S and SOL in what follows, but exactly the same points carry over to $Q_{2^{\aleph_0}} x(x = x) \leftrightarrow Q_{\aleph_1} x(x = x)$ and $\mathcal{L}_{\infty\infty}$.

Writers such as Feferman and Bonnay aim to jump from biconditionals such as 'S is logically true iff CH is true' to a problematic conclusion about CH. Exactly what this problematic conclusion is differs: Feferman claims that CH is rendered 'logically determinate' (Feferman 2010a, p. 12) by the isomorphism-invariance approach; Bonnay claims that it is 'possible to express' (Bonnay 2008, p. 37) CH using isomorphism-invariant vocabulary; and as we saw earlier in this section, other writers put the point in different ways as well.

---

[4] See e.g. Shapiro (1991, §5.1.2) for the details.

[5] In slightly more usual notation, this may simply be written Q_α. The notation $Q_{\aleph_\alpha}$ is preferable here because of its greater transparency.

[6] For the definition of sort, see Chapter 4.

Now, whether a sentence is true and contains only logical vocabulary on the one hand and whether it is logically true on the other are independent questions. The table contains examples of all four categories:

	Only logical constants	Some non-logical constants
Logically true	If something exists then it exists.	If a planet exists then a planet exists.
True but not logically true	There are at least 8 things.	There are at least 8 planets.

So if their objection is to stick, Feferman, Bonnay, and company must first show that isomorphism invariance implies that the sentence S is logically true if CH is true.[7] And as we saw in Part II more generally, deriving a conclusion about the one true logic (and hence about which truths exactly are logical) from an account of logical constants involves a good deal of supplementary work and assumptions. But since it is in line with our broader conclusions in Part II, we shall grant critics of isomorphism invariance this premise and assume that second-order logic or $\mathcal{L}_{\infty\infty}$ is (at least part of) logic. The question now is what the problematic conclusion about CH is supposed to be.

Unfortunately, the critics are not very clear on this point either, which is why we shall have to do much work on their behalf. In §§9.3–9.7, we shall consider several possible overgeneration objections that may be raised against a position such as ours, for example that it renders CH either *logically* true or false, or that it renders CH epistemically determinate. We will explain why none of the precise versions of the overgeneration objection we consider succeeds. Of course, it is open to the critics to put forward another reading, but we shall cover anything that might be teased out of the existing literature and it is not at all obvious what a plausible alternative reading would look like.

In all of these arguments, it is the status of CH, rather than the isomorphism-invariance test, that is crucial: if you accept the second-order quantifiers (with full semantics) as logical for other reasons than isomorphism invariance, you should still be concerned by the conclusions of these arguments. For this reason, many of our responses have a wider target: they should move not only the opponent of isomorphism invariance but also anyone who believes that the logical status of second-order quantification is problematic. When our arguments in §§9.3–9.7 are directly aimed at the opponent of isomorphism invariance we will be explicit; otherwise, they are aimed at this broader group. First, we start in §9.2 with the general and vague idea that logic, if it encompasses second-order logic or $\mathcal{L}_{\infty\infty}$ (or both, as explored in Chapter 7), is sensitive to mathematics. We introduce this

---

[7] And that S' is logically true if CH is false, and similarly for analogous interpreted sentences of $\mathcal{L}_{\infty\infty}$. As we said, we usually take these elaborations as read.

idea in §9.2; working on our critics' behalf, we then look at ways of making it more precise in §§9.3–9.7.[8]

9.2 Sensitivity

The critic of isomorphism invariance believes that the CH example reveals a certain *intimacy* between mathematics and logic when isomorphism invariance is taken as our criterion of the logical. The thought is that isomorphism invariance renders logical truth *sensitive* to set theory.

The critic of isomorphism invariance must do more to spell out what 'sensitivity' amounts to in this context. For this reason, we will do much work on our opponents' behalf by providing the four most plausible precisifications of 'sensitivity'. In this section, we work with the intuitive notion of 'sensitivity' that is often appealed to by critics of isomorphism invariance. The discussion in the five sections to follow (§§9.3–9.7) attempts to unpack this criticism.

Sticking to the rather vague notion of sensitivity, at least a minimal reading must imply that the extension of 'logical truth' varies with the truth-value of mathematical sentences: if CH is true, then S is a logical truth; if CH is false, then S' is a logical truth. This sensitivity to the truth-value of controversial set-theoretic claims, the critic may argue, is reason enough to be suspicious of isomorphism invariance.

The full moral about how to distinguish a set-theoretic from a logical claim will emerge later, but can already be exemplified by a brief comparison between CH and S. CH is a mathematical sentence, which is not topic-neutral, since it is about sets. Its truth is existentially committing: it requires the existence of the set of functions from $\aleph_0$ to 2, the cardinal $\aleph_1$, and of a bijection between them. CH is therefore not *logical*, since it is existentially committing and has mathematical content.

Turn now to S, defined in §9.1. There are different ways to interpret the second-order quantifiers: as quantifying over properties, or pluralities, or still other types of collections. If we interpret them set-theoretically, S will of course be in some sense about sets. But there will be nothing special about S—which critics see as problematically tangled up with CH—in this regard, since *any* interpreted second-order sentence with second-order quantifiers will by the same token be about sets. Consider for example $\forall XXb$ thus interpreted over a domain of all the horses past, present, and future, the constant b denoting Bucephalus. The resulting interpretation is (the false) 'Bucephalus is a member of all sets of horses'. Proponents of the overgeneration objection clearly do not take themselves to be

---

[8] Our discussion in this chapter builds on two previous articles: Paseau (2014) and, especially, Griffiths and Paseau (2016). These were in turn influenced by Gómez-Torrente (1998) and Soames (1999).

merely making this Quinean point,[9] which applies to any old sentence of second-order logic proper.

To try to get a grip on what precisely the problem with S being a logical truth is supposed to be, it will be helpful to take the interpretation of the quantifiers to *not* be set-theoretic, so that S does not make any mention of sets. We may, for example, interpret the quantifiers as plural quantifiers (as pioneered in Boolos (1984)), so that S is the statement

> For any things$_X$, things$_Y$ and things$_Z$, if there are as many things$_X$ as natural numbers and as many things$_Y$ as real numbers and there are fewer things$_X$ than things$_Z$, then the things$_Y$ are fewer or equal to the things$_Z$.

(This sentence is of course a mouthful, owing to its complexity.) Since the overgeneration point, as explained, is supposed to be distinct from the Quinean worry that any interpretation of the second-order quantifiers must ultimately be set-theoretic, it is clearest to assume that some non-set-theoretic interpretation (such as the plural one) is available and that S results from interpreting the formal sentence $\forall X \forall Y \forall Z((N(X) \wedge R(Y) \wedge X < Z) \rightarrow Y \leq Z)$ in this way. Otherwise, we might risk confusing the Quinean point (that any interpretation of the second-order quantifiers must invoke set theory) with the overgeneration objection (that the connection between S's logical truth and CH shows that logic is problematically sensitive to mathematics).

Returning to the comparison between CH and S, we have observed that CH is a mathematical sentence. S, on the other hand, is not a mathematical sentence. It is topic-neutral, since it can be expressed using only logical vocabulary and has no special subject matter. Nor is it existentially committing. Florio and Incurvati (2019) say that one might take issue with S's ontological neutrality, either for the broadly Quinean reason that non-logical notions are hidden behind second-order ones or, more generally, because one might take second-order quantifiers to carry ontological commitment. But S has the form of a conditional prefaced by three leading universal quantifiers, so it is not existentially committing because its truth does not require the existence of any entities, of any order. In the same way, the sentence 'all numbers, sets, and vector spaces are such that if...then...' is not existentially committing, because it is true if there are no numbers, sets, or vector spaces. Finally, note that S contains only logical expressions.[10]

CH and S are, then, very different sentences and respecting these differences will be crucial in what follows.[11] The reason to suspect that the general 'sensitivity'

[9] See Chapter 5 of Quine (1970).

[10] On the assumption that SOL is part of OTL, which we are assuming here, as explained.

[11] We have favoured the existential reading of CH here, but of course relative to ZFC that existential version is equivalent to a universal one much more similar to S. Our point here is to highlight the difference between a mathematical claim such as CH with its usual existential interpretation and a claim such as S; it is not to suggest that CH must always be formalized in this way.

objection is mistaken can be brought out by analogy with an arithmetical case. Consider the sentences

A   2+5=7

A′ $(\exists_2 xFx \land \exists_5 xGx \land \forall x \neg(Fx \land Gx)) \rightarrow \exists_7 x(Fx \lor Gx)$

Here A and $A′$ are both interpreted,[12] $\exists_2 xFx$ abbreviates $\exists x \exists y(Fx \land Fy \land x \neq y \land \forall z(Fz \rightarrow (z = x \lor z = y)))$, and likewise for $\exists_5 xFx$ and $\exists_7 xFx$. A is a truth of arithmetic which is about numbers, hence not topic-neutral nor, by our lights, logical; those who take first-order logic to be logic would agree since A's first-order formalization is $f(a, b) = c$. $A′$ is, however, a first-order logical truth. The situation is analogous with the case of CH: just as S is logically true iff CH is true, $A′$ is logically true iff A is true. The only disanalogy is that the mathematical claim CH is controversial, whereas everyone accepts the truth of A.[13]

The logical truth of $A′$ is sensitive to the truth of A in just the same way as the logical truth of S is sensitive to the truth of CH: again, the extension of 'logical truth' will differ depending on the truth-value of a mathematical claim. The logical truth of both sentences S and $A′$ is equivalent to the truth of their corresponding mathematical sentences. And yet this is not thought to be problematic in the first-order case. Now if the sort of sensitivity under discussion is sufficient for overgeneration, then there is overgeneration at the first-order level, which critics of isomorphism invariance typically deny. Therefore, isomorphism invariance does not render the extension of 'logical truth' hostage to mathematics any more than is standard for first-order logic.

One could respond by denying the logical nature of identity, but we will defer discussion of this response until §9.8. It should be equally clear from our discussion that isomorphism invariance does not '[require] the existence of set-theoretical entities of a special kind, or at least of their determinate properties' (Feferman 2010a, p. 9). As above, CH (i.e. $2^{\aleph_0} = \aleph_1$) requires the existence of sets, namely the set of functions from $\aleph_0$ to 2, $\aleph_1$ and the existence of a bijection between them; similarly for ¬CH ($2^{\aleph_0} \neq \aleph_1$), minus the existence of the bijection. In contrast, neither S nor $S′$ is committed to the existence of a single set, nor *a fortiori* to any sets having determinate properties.

Another response could be that the arithmetic A and the set-theoretic CH are very different. A, this response would go, is much simpler and more familiar than CH, so logic's sensitivity to A is less problematic than to CH. This response is not a straw man: Bonnay (2008, p. 56) distinguishes *problematic* from *unproblematic* mathematical content. We believe that this is an unstable distinction. There are

---

[12] In interpreting $A′$, we interpret F and G as distinct properties/pluralities—which ones exactly does not matter.

[13] Bracketing a few fictionalists, formalists, and the like.

two main ways to understand it: that set-theoretic entities are *ontologically* more problematic than arithmetic ones, or that they are *epistemologically* more problematic. As far as the ontological point goes, numbers and sets are as good as each other or as bad as each other, or so it seems. Overgeneration critics think that logic should not be sensitive to mathematics, and in particular to neither numbers nor sets. They might be worried that 'infinitary' structures behave differently from 'finitary' ones, in a way that is very relevant to the nature of logic; but if so, that argument remains to be made. On the epistemological side, numbers and sets raise similar access problems. Furthermore, although the truth of A is obvious, there are first-order logical truths whose truth is far from obvious and recognized only on the basis of sophisticated mathematics (§9.5 develops an example involving primality).

Our diagnosis for why someone might think S but not A' problematically mathematical is that we tend to be much more familiar with first-order logic. Both A' and S are logical truths that have mathematical counterparts, one of which is obviously true and the other far from it, just like A' and S respectively. (On the assumption that CH is true.) This verdict will be borne out when we examine more precise versions of the sensitivity argument.

9.3 CH is *logically true*

We will now consider some precisifications of the thought that logic is *sensitive* to mathematics on the isomorphism-invariance account. In this section, we consider a constitutive reading: CH is rendered *logically* true, if true; and *logically* false, if false. This would be a problematic conclusion for isomorphism invariance, since we motivated it with the thought that logic is topic-neutral and CH should therefore not count as logical.

There is an interesting parallel here with Etchemendy's arguments against the model-theoretic definition of logical consequence (1990, pp. 123–4). He maintains that the model-theoretic definition is either committed to the claim that CH is logically true, or the claim that CH is logically false. It is not at all clear how Etchemendy's argument is meant to work here (Paseau 2014). For our purposes, it will be helpful to adapt one of the precisifications canvassed in Paseau (2014).

CH is a claim made in (a technical fragment of) natural language, which if fully formalized in first-order terms would be a very long expression in the first-order language of set theory containing a single dyadic predicate $\in$. As noted, it is also existentially committing since, for example, $2^{\aleph_0} = \aleph_1$ entails the existence of the cardinals 2, $\aleph_0$ and $\aleph_1$. The sentence that is logically true if CH is true is rather the interpreted second-order sentence S, which, recall, is:

$$S \quad \forall X \forall Y \forall Z((N(X) \wedge R(Y) \wedge X < Z) \rightarrow (Y \leq Z)).$$

But the logical truth of *S* is not problematic, assuming CH is true. Consider again an analogy with an arithmetical pair derived from the earlier *A* and *A'*.

A   2+5=7

A'' $\forall F \forall G((\exists_2 x Fx \wedge \exists_5 x Gx \wedge \forall x \neg (Fx \wedge Gx)) \rightarrow \exists_7 x (Fx \vee Gx))$

As with *A'*, *A''* is by stipulation an interpreted sentence. *A* is a truth of arithmetic about numbers which is not topic-neutral and therefore not a logical truth. *A''* is, however, a second-order logical truth. The situation is as discussed in §9.2: just as *S* is logically true iff CH is true, *A''* is logically true iff *A* is true. So if *A* is true, the second-order sentence *A''* is logically true. But the logical truth of *A''* is not problematic, since it is topic-neutral. For example, a natural way to express *A''* in English might begin 'for any two properties, say *F*-ness and *G*-ness, if there are exactly two *F* things and exactly three *G* things, . . .'. *A''* is therefore a topic-neutral sentence and its logical truth is unproblematic. And similarly for *S*.[14] The crucial point is that neither of the mathematical sentences CH or *A* is declared logically true. The fact that they are materially or metaphysically equivalent to logical truths is not enough for them to be logical truths themselves.

To be clear about the dialectic, we are not here *assuming* that second-order quantifiers are logical expressions. There is a well-known debate about the second-order quantifiers that is quite independent of isomorphism invariance. For example, if you worry about the logical status of monadic second-order quantification, or its topic-neutrality, you may offer a plural interpretation, as mentioned. We are not assuming any such justification of the second-order quantifiers. Our argument runs in precisely the opposite direction: isomorphism invariance is motivated by the thoughts that the logical constants are topic-neutral and general. The isomorphism invariance of second-order quantifiers shows that they are topic-neutral and general in the same way as first-order quantifiers. In this way, with further work to connect the right account of logical extensions with the right account of logic (see Part II), isomorphism invariance can be used to defend second-order logic as logic. This defence dovetails with pluralist justifications of second-order quantification as logical but does not presuppose them. What would be problematic is if an *obviously* non-topic-neutral

---

[14] Observe that, in each case, the logical sentence (respectively *S* and *A''*) can be derived from the mathematical one (respectively CH and *A*) in *roughly* the following manner: turn the names for cardinals (e.g. the name for $2^{\aleph_0}$ in CH and the name for 2 in *A*) into predicate variables whose instances are then specified to have the appropriate numerosity; turn the identity symbol into a conditional; universally quantify over the new predicate variables; and, modulo a couple of further transformations, one obtains the logical sentence (*S* and *A''* respectively). The differences between the two procedures have mainly to do with the fact that the left-hand side of CH is an exponentiation whereas the left-hand side of *A* is a sum.

sentence like CH were deemed logically true, if true. And we have shown that it is not.

Another way to argue that CH is rendered logically true on the isomorphism-invariance account is to consider the behaviour of the 'it is a logical truth that' operator.[15] If CH is true then so is the following :

(1) It is a logical truth that S.

This, coupled with

(2) $S \leftrightarrow$ CH

is not yet sufficient to entail:

(3) It is a logical truth that CH

since the entailment would involve a modal fallacy on any reasonable account of the logic of logical necessity.[16] However, we can amend (2) in the following way:

(2′) It is a logical truth that: $S \leftrightarrow$ CH.

(1) and (2′) *do* entail (3).

The problem for our opponents is now twofold. First, they must argue for (2′), which we see no reason to believe. Second, such an argument is not available to our opponents because it proves too much. To return to the arithmetic analogy, A'' is a logical truth so the analogous conclusion here would be that the arithmetical sentence A is a logical truth. But A is not a logical truth, since it is not topic-neutral. Indeed, to accept A as a logical truth would be to endorse a form of logicism. But logicism is explicitly rejected by at least one of our main opponents (see Feferman 1999), and faces numerous well-known objections.

9.4 CH is *determinate*

Opponents of isomorphism invariance, such as Feferman in particular, could put a different spin on the sensitivity objection. They could argue that logical determinacy entails a problematic mathematical determinacy. As background,

---

[15] Paseau (2014, p. 46) considers this sort of argument in his response to Etchemendy.

[16] That this is a fallacy is agreed on all hands. For discussion of the logic of the 'It is a logical truth that' operator and the related 'It is true in virtue of logical form that' operator, see Paseau and Griffiths (2021a).

we briefly describe Feferman's unusual views about set theory. Along with most others, Feferman maintains that standard iterative set theory describes a unique universe of sets. The iterative universe of sets is build up by various operations, notably by taking the power set of the set of all sets previously formed at any successor stage. Where Feferman's idiosyncratic understanding of the set universe comes in is that he takes the power set operation to be inherently indeterminate when applied to infinite sets. This leads Feferman to view the entire set universe as indeterminate (though unique). In particular, he takes the truth-value of a statement such as CH (which concerns the size of the power set of ω) to be indeterminate. If we then take logic to include second-order logic and view logic as determinate, this combination of views leads to trouble.[17]

To formalize the argument, let $\Box_M$ and $\Box_L$ express mathematical necessity and logical necessity respectively. First, suppose you hold that logic is determinate, and in particular that S is either determinately logically true or that its negation (which we earlier called S' but will here call $\neg S$) is determinately logically true.

(1) $\Box_L S \lor \Box_L \neg S$

Second, logical truths are mathematically necessary. Indeed, logical truths are completely topic-neutral, whereas mathematical truths may rely on the existence of particular objects. So we have the schema

(2) $\Box_L \phi \to \Box_M \phi$

and its two instances

(2a) $\Box_L S \to \Box_M S$
(2b) $\Box_L \neg S \to \Box_M \neg S$

Third, it is plausible that 'CH $\leftrightarrow S$' is *mathematically* necessary. It is certainly not *logically* necessary, as above, since the truth of CH requires the existence of sets that do not exist of logical necessity, but let's allow that it's mathematically necessary.[18] Hence:

(3) $\Box_M (\text{CH} \leftrightarrow S)$

---

[17] For Feferman's views, see his (2011). See also Feferman (2010b) for more technical discussion and Scambler (2020) for a development of Feferman's ideas. We are grateful to Tim Button for a formalization of Feferman's argument similar to the one that follows, and for helpful discussion of Feferman.

[18] If you doubt that 'CH $\leftrightarrow S$' is mathematically necessary, then this version of the overgeneration argument doesn't even get off the ground.

(1), (2a), (2b), and (3) entail:[19]

(4) $\Box_M$ CH $\lor \Box_M \neg$ CH

To summarize the objection: *if* the logical truth of *S* is determinate, and we accept the plausible claims (2) and (3), then the *mathematical* truth of *CH* is determinate. And this is a commitment that someone like Feferman, who believes the universe of sets is indeterminate, will not accept.

As an aside, we note that a proponent of this argument must cash out the notion of *mathematical* necessity on which it crucially depends. One explication of this idea that immediately suggests itself is the set-theoretic multiverse view of, for instance, Hamkins (2012). This view, very roughly, is that there are many different conceptions of set, each of which is instantiated in a corresponding set-theoretic universe. The view provides us with a nice explication of the notion of mathematical necessity, since we can interpret $\Box_M \phi$ as 'ϕ is true in all set-theoretic universes in the multiverse'. Such multiverse views are, of course, controversial, and we do not accept them (see Chapter 10). Nor does it seem that Feferman does, since his position is that there is a unique, though indeterminate, universe of sets. But for the sake of argument, we may assume that some account of mathematical necessity is available.

Our response to the argument is to ask whom it is supposed to affect. Suppose that, like us, you believe that *S* is logically determinate and CH is mathematically determinate. Then you will not be troubled by the argument, as you embrace its conclusion. This is what virtually everyone believes. If that includes you, there is nothing here to be troubled by.

Suppose alternatively that, like Feferman, you believe that the truth of CH *is* indeterminate. Then it is hard to see how you could accept premise (1). On the standard model-theoretic definition of logical truth, a sentence is a logical truth just if it is true in all domains under all reinterpretations of the non-invariant vocabulary. But if you believe that there are indeterminate sentences about the background universe of sets, then it is unsurprising that some of this indeterminacy carries over to logical truth. If you hold that CH is indeterminate, you will think that there are sentences whose logical status is indeterminate. The hybrid position that sees the mathematical sentence CH as indeterminate and the logical truth of the second-order sentence *S* as determinate is ill-motivated. Of course, there are other accounts of logical truth, in particular inferentialist views; but on a typical inferentialist view, neither *S* nor its negation is a logical truth, since inferentialists typically do not accept that second-order logic (with standard semantics) is logic.

---

[19] Via propositional logic and Distribution for $\Box_M$.

In sum, the thought that the truth of CH is mathematically determinate is much the most usual one, which we accept. If you incline to the controversial view that rejects it, you will and should also reject the argument's first premise. Either way, the connection between logical determinacy and mathematical determinacy has not been shown to be problematic.

9.5 CH is *epistemically determined*

We now turn to our next reading of *sensitivity*: CH is *epistemically determined* by isomorphism invariance. The thought here is that logic alone should not force us to accept or deny controversial claims of set theory, so isomorphism invariance is not the correct criterion for logical nature.

First, it will be instructive to rehearse the argument of §9.3, but with the operator 'it is an *a priori* truth that' in place of 'It is a *logical* truth that':

(1) It is an *a priori* truth that S.

This, coupled with

(2) It is an *a priori* truth that: $S \leftrightarrow$ CH.

entails

(3) It is an *a priori* truth that CH.

Given Distribution for 'it is an *a priori* truth that', this argument is valid. But its conclusion should not worry us, or at least it should not worry anyone inclined to think that S is knowable *a priori*. And if CH is true and knowable, then presumably it is knowable *a priori*. What should worry us is the *logical* truth of CH and we have not seen any sound argument for that conclusion.

Returning to the criticism that CH is epistemically determined by isomorphism invariance, we accept, of course, that CH is an unsettled claim of set theory. What we deny is the conditional claim that if isomorphism invariance is the correct criterion for logical nature, then some new and problematic epistemic route to knowledge of CH or of ¬CH, as the case may be, becomes available.

To begin with, let us be clear about the expected order of discovery. The criticism seems to imply that we might first discover that S is a logical truth and thereby infer the truth of CH. But there is another sentence S′, which is logically true just if CH is false, and S′ also contains only isomorphism-invariant vocabulary. So isomorphism invariance as a criterion for logical constanthood is consistent with both the truth of CH (in which case S is logically true and S′ is false) and the falsity of CH (in which case S' is logically true and S is false). In short, it is not our account in Part II that delivers the logical truth of S: it is that *and* the truth-value of CH.

If we ever settle CH's truth-value, the expected order of discovery is rather that we shall *first* determine the truth of CH via set-theoretic techniques—from mathematically well-motivated extensions of ZFC—and *then* infer the logical truth of S (or S'). The proponent of isomorphism invariance is in no way committed to, indeed rejects, the idea that 'a competent speaker should be able, on the basis of their semantic competence, to accept or reject the Continuum Hypothesis'.[20] Even in the first-order case, in which there is a sound and complete proof procedure, the order of discovery may very well run from mathematics to logic. For example, there is a statement ϕ_N of first-order logic which is logically valid (true in all interpretations) iff the arithmetical statement 'N is prime' is true. The statement ϕ_N is the negation of

$$\bigvee_{\substack{1<b<N \\ 1<a<N}} Mult(a,b,N),$$

a disjunction which consists of $(N-2)^2$ disjuncts. $Mult(a,b,N)$ is the first-order formalization of the following statement: If there are exactly a Fs, and each F is R-related to exactly b Gs, and no two distinct Fs are ever R-related to the same G, and any G is the R-relatum of some F, then there are exactly N Gs. (Intuitively, $Mult(a,b,N)$ 'says' that $a \times b = N$.) Advances in, say, analytic number theory may determine that a particular number N is prime, thereby yielding knowledge that ϕ_N is a logically valid formal sentence (so that any interpretation of ϕ_N is logically true).[21]

Second, exploiting logical knowledge to yield mathematical knowledge should not be regarded as problematic. Again, this should be familiar enough from the first-order case, for instance, our 'N is prime' example. From knowledge of the material biconditional that 'N is prime' is true iff ϕ_N is a first-order validity, combined with logical knowledge of the fact that ϕ_N is a first-order validity (say via a premise-free first-order proof), one may come to know that N is prime. Coming to know that CH is true analogously, via knowledge of S and the biconditional $S \leftrightarrow$ CH, is no more problematic than coming to know that N is prime by exploiting knowledge of the logical truth ϕ_N. As mentioned in the previous paragraph, the converse is also true: we can and do exploit mathematical knowledge to yield knowledge of logical truth.

Third, and related to the second point: that *some* logical knowledge has been exploited to yield mathematical knowledge does not mean that *only* logical

---

[20] Bonnay (2008, p. 65), with the original 'its' replaced by 'their'.

[21] Analytic number theory considers the natural numbers as embedded in the complex plane and uses techniques from complex analysis to derive facts about them, in particular facts about primes. The canonical proof of the Prime Number Theorem is a famous case in point.

knowledge has been exploited. In our arithmetical example, *logical* knowledge of the primality of N only follows from logical knowledge that ϕ_N is a logical validity and *logical* knowledge of the material biconditional that 'N is prime' is true iff ϕ_N is a logical validity. Suppose, as we believe, that the biconditional is *not* logically knowable. Then this route to knowledge of the primality of N is not purely logical, as it rests on extra-logical knowledge of the biconditional. Similarly, coming to know CH via presumed logical knowledge of S and knowledge of the biconditional 'CH iff S' yields logical knowledge of CH only on the assumption that the biconditional is logically known. Yet we see no reason to suppose that this biconditional *is* logically knowable.

In sum, knowledge of S may in principle lead to knowledge of CH, although the order of discovery is almost bound to run in the other direction. As for logical knowledge of S supposedly leading to logical knowledge of CH, not only has the case for this supposition not been made, it is also highly problematic.

9.6 CH and neutrality

In a recent paper, Florio and Incurvati (2019) have adduced two further ways in which to understand the 'entanglement' of logic and mathematics brought about by isomorphism invariance, indeed by mere acceptance of second-order logic as logic. Their arguments are based on the respective ideas that logic should be *dialectically* and *informationally* neutral. If it fails to be neutral in either of these ways, then it is problematically 'entangled'. Let's take these two forms of neutrality in turn.

9.6.1 Dialectical neutrality

Starting with dialectical neutrality, what Florio and Incurvati mean by this is that logic should not take sides in debates in mathematics, science, or metaphysics. Instead, logic should be a neutral arbiter. Yet, as they see it, the fact that S is a logical truth iff CH is true shows that logic, thus conceived, is not dialectically neutral. This is another way of saying that it overgenerates.

We take the doctrine that logic should be dialectically neutral to be highly controversial. It is easy to list examples of logics put forward on various metaphysical grounds, from fuzzy logic to various modal logics, from intuitionistic logic to paraconsistent logic, and from quantum logic to many-valued logic. Whether the debate is about the future, the infinite, paradoxes, possible entities, physical phenomena, or vagueness, many of these logics' proponents have no intention to adopt a 'neutral' stance. Of course, some proponents of weakening classical logic do appeal to neutrality: 'even if this metaphysical view is false, its falsity should not be written into logic', they might say. But they in turn neglect metaphysical challenges to their preferred weaker logic, as Williamson (2014) points out in a robust challenge to the neutrality of logic.

Furthermore, requiring that 'logic should be neutral' does not much help adjudicate these debates. Even when all participants to a debate or dispute agree to subject themselves to a neutral arbiter, it is extremely hard to circumscribe the neutral arbiter's powers in a neutral way. International politics sadly provides plenty of evidence of the 'second-order problem of neutrality'—how to find a neutral arbiter in a neutral way. In the field of logic, an illustration is the traditional debate between classical logicians and pretenders to the throne. The former happily look on first-order logic as dialectically neutral. But of course from the pretender's point of view, classical logic imports some subject-specific and 'partisan'—non-neutral—principles. Intuitionists are a case in point: they take the unrestricted form of the Law of Excluded Middle to be a mathematical assumption when applied to mathematics, and similarly in other domains.

In short, invoking the dialectical neutrality of logic in order to uphold the overgeneration objection seems to us unpromising. A relatively concrete issue about the 'equivalence' of S and CH has turned into a much larger debate about a contestable—and much-contested—trait of logic. Invoking dialectical neutrality is itself dialectically ineffective.

In fact, it is far from clear that the (suspicious) requirement of dialectical neutrality even sustains the overgeneration objection. In the case of SOL, specifically, one can imagine a defence of its dialectical neutrality proceeding as follows. SOL itself *is* dialectically neutral, because its interpreted logical validities have no mathematical content, when this latter notion is precisely understood. They make no existence claims about sets or any other mathematical objects, for example. Such claims can be extracted from SOL in combination with 'bridge principles' of the kind linking S and CH, for example. But make no mistake: these bridge principles are inherently mathematical, as we observed above. By itself, SOL is dialectically neutral; it is only SOL plus a host of logical-to-mathematical bridge principles—which are *mathematical* rather than logical—that is not neutral. And it's hard to see what dialectical neutrality in the present mathematical context even means if we are allowed to assume, say, ZFC, to prove the bridge principles. In fact, given their own proposed solution to the overgeneration problem (whose summary follows towards the end of this section), Florio and Incurvati are likely to be sympathetic to this response.

We conclude with two final points on dialectical neutrality. Whether a logic is dialectically neutral or not is an epistemological matter. But, as noted in §9.5, our only current means of resolving issues such as CH are mathematical. The idea that someone might come up with a non-mathematical way to establish that S, say, is a logical truth and *thereby* establish CH is hard to imagine. As far as dialectical neutrality *in practice* goes, then, logic *is* neutral: we cannot use it to settle any interesting mathematical claims. Our second point is that an analogy may once again be drawn between CH and S, on the one hand, and A, and A', on the other (see §9.2). If you are worried about SOL's dialectical neutrality because its logical truths settle *CH* in the presence of appropriate bridge principles, you should by the same token be worried about FOL's dialectical neutrality because its logical truths settle arithmetical statements in the presence of appropriate bridge principles. Our recommendation is to perform a modus tollens on this argument and not to worry about SOL's dialectical neutrality.

9.6.2 Informational neutrality

Let's turn now to Florio and Incurvati's other way of cashing out the overgeneration objection, by means of informational neutrality. The basic idea, roughly put, is that informational content is preserved by logical consequence. Second-order theories, however, seem to spin informationally rich consequences out of premises that are informationally meagre. Florio and Incurvati cite the example of Robinson Arithmetic, whose natural second-order formulation turns it into second-order Peano Arithmetic, a much stronger theory. Logic, however, cannot bring about an increase in informational content all by itself. So second-order logic overgenerates, informationally.

This form of the objection is not specific to cases like the example of S and CH. As their example of arithmetic shows, it is a more general worry that SOL is 'too strong'—it generates too many consequences from given premises. The case of S and its link to CH is just one way of illustrating the point. It therefore seems a quite distinct and much more generic sort of worry than the overgeneration objection as understood by writers such as Etchemendy, Blanchette, Priest, Shapiro, Feferman, and Bonnay (see the earlier quotations).

In any case, exegesis aside, it is unfortunate that Florio and Incurvati do not define more precisely the notion of informational content they rely on. In its absence, cashing out the somewhat inchoate overgeneration worry in terms of informational surplus cannot be considered a major clarification of the issue. Left at an informal and intuitive level, the criterion of informational neutrality seems to clash with all but the most impoverished conception of logic. Dialetheists and other paraconsistentists aside, logicians take a contradiction such as 'It's noon and it's not noon' to entail any statement whatsoever: that Fermat's Last Theorem is true, that the Moon is made of green cheese, that Homer wrote the *Odyssey*, and so on. But does any contradiction really *contain* all statements' informational content? Is it informationally maximal? Similarly, do any two propositionally equivalent statements have the same informational content? For instance, do tautologies such as 'It's noon or it's not noon' and 'If Fermat's Last Theorem is true then it's true' have exactly the same content? Does the informational content of both include the claim that Paris is in France iff it's in France? It would be easy to multiply counterintuitive consequences of Florio and Incurvati's credo—that logic cannot generate 'extra' informational content—if informational content is understood in a vague, informal way.

Can the objection be sharpened by appeal to some precise notion of informational content? Although it is up to Florio and Incurvati to provide such a notion, so far as we can see none fits the bill. A proponent of SOL could lay down the definition

Statement s_1 informationally contains s_2 iff s_2 is an SOL-consequence of s_1.

But of course that would collapse the overgeneration objection, not buttress it. The reason is that, following the definition, any premise that SOL-entails a conclusion has not generated any extra content, since the premise's information content contains the conclusion's informational content. More generally, the proponent of $\mathcal{L}$ as the one true logic is likely to understand 'informational content' in the same way, replacing SOL with $\mathcal{L}$ so that

Statement s_1 informationally contains s_2 iff s_2 is an $\mathcal{L}$-consequence of s_1.

It would then follow, as for the second-order case, that $\mathcal{L}$ is informationally neutral. For if a premise $\mathcal{L}$-entails a conclusion then, by this criterion, the conclusion's informational content is part of the premise's.[22]

Alternatively, one could take informational content as truth-conditional content, so that two statements have the same informational content iff they have the same truth-conditions. This does not help either unless we are told what truth-conditions are. In fact, on perhaps the most popular way of understanding them, as sets of metaphysically possible worlds, S and CH are truth-conditionally equivalent.

A third and more fine-grained approach would be to read informational content as identity of Fregean sense. On this construal, 'I respect Dummett' and 'Dummett is respected by me' have the same informational content, but S and CH clearly do not (e.g. the former is a universal claim whereas the latter is an existential claim that posits the existence of a certain bijection). The problem with this construal is the same as that for the informal intuitive one: informational content is not preserved by all but the weakest notion of logical consequence. For example, it's a consequence of 'I respect Dummett' that 'I respect Dummett or the Moon is made of green cheese', but the latter statement's sense is not contained in the former's.

In any case, as we mentioned earlier, we are on the same side as Florio and Incurvati. Their claim is merely that SOL *appears* to overgenerate because it *appears* not to respect informational neutrality. They try to dispel that impression by

---

[22] Similarly for premise sets.

showing that S and CH are not interderivable in a weaker metatheory than those on the basis of which the overgeneration objection is usually made. This is the theory they call ZFC*. We are quite clear ourselves that S and CH are not informationally equivalent in any problematic sense. Whether Florio and Incurvati can motivate their account of informational content, and then demonstrate that logic must be informationally neutral in something like this precise sense, is a further question. The success of our defence does not defend on the success of theirs. We doubt that invoking informational content will be helpful here unless it is a label for some other condition shorn of informational content's intuitive properties. But if they do succeed, it will be in the service of reaching the same sort of conclusion as us: S and CH are not 'equivalent' in any problematic sense. We have a little more to say about Florio and Incurvati's proposal, but since nothing in this book rests on it, we relegate it to a footnote.[23]

Moving on from neutrality, dialectical and informational, let's pause briefly to summarize where we have got to. The last four sections considered more precise versions of the criticism that isomorphism invariance renders logic sensitive to mathematics. The inchoate worry that second-order logic is 'entangled' with mathematics cannot, we believe, be cashed out in a way that is both more precise and still troubling. Rebutting every possible precisification of the worry would be

---

[23] Florio and Incurvati claim that if we move from set-theoretic to higher-order semantics for the second-order quantifiers, we can no longer establish the equivalence of CH and S. They define ZFC* to be the set of sentences derivable from ZFC in their axiomatization of SOL, which includes the usual rules for the second-order quantifiers, as well as all instances of the usual Comprehension Scheme for SOL. For the record, this is: $\exists X^n \forall x_1 \ldots \forall x_n (X^n x_1 \ldots x_n \leftrightarrow \phi(x_1 \ldots x_n))$. In ZFC*, one cannot prove that CH implies S nor the converse. The set-theoretic truth CH is no longer 'entangled' with the logical truth S.

As mentioned, Florio and Incurvati are on the same side as us: they too aim to exonerate SOL of overgeneration. Since we take logic to be at least second-order (as well as maximally infinitary), their result could in principle help us too. Observe that, intuitively, their result is just what one would expect. If you restrict ourself to a metatheory in which there are no or only impoverished 'bridge principles' between sets and the higher-order entities over which the second-order quantifiers range, you would expect claims about sets to be largely independent of claims involving quantification over higher-order entities. As Florio and Incurvati note, as soon as you add such bridge principles—in particular, the second-order version of Replacement—the equivalences are resurrected. As they equally acknowledge, one of the best-known proponents of higher-order semantics—Timothy Williamson (2014)—repudiates the idea of logic as neutral. Even friends of the Florio–Incurvati approach, then, recognize that it is *not* part and parcel of the higher-order semanticist's approach that logic should be neutral and that any principles that imply equivalences between mathematical and logical statements are thereby suspect.

To evaluate Florio and Incurvati's response to their overgeneration objection, then, one would have to assess their proposed higher-order semantics very carefully. As we have seen, severing the S–CH equivalence delicately depends on the nature of the higher-order semantics and the principles it assumes. We also need a story about why the mathematical equivalence of S and CH (given mathematical bridge principles linking mathematics to logic) is *more* problematic than mathematical equivalence of A and A'. The worry seems to be that second-order logic is problematic because of the degree to which it is entangled, not the fact that it is entangled. Yet against this, we have argued (a) that it's not clear why a moderate amount of entanglement is any better than a large amount; and (b) that there is no entanglement in the first place. In any case, although we do not think that the move to higher-order semantics is required to defeat the overgeneration objection, such a response is now also on the table, thanks to Florio and Incurvati.

an impossible task. Like the Lernaean Hydra's heads, the argument seems to sprout more forms as soon as existing ones have been cut off. But we take what we have said in this chapter as a reliable recipe for responding to future forms.

9.7 CH is logically *expressible*

We turn finally and more briefly to Bonnay's claim that it is possible to express CH and other complex set-theoretic claims using only isomorphism-invariant expressions (Bonnay 2008, p. 38). We can now see that this objection is misguided: what is expressible in purely logical vocabulary is not CH but S. And if, as we have argued, the logical truth of S is unproblematic, then of course its expressibility is unproblematic, since a sentence's logical truth (in a particular logic) implies its expressibility (in that logic). If a sentence is a logical truth of second-order logic, then of course it can be expressed in the language of second-order logic. And if the former should not trouble us, nor should the latter.

Now mathematicians typically do not make fine distinctions between state-ments expressing what they might regard as the same 'mathematical content'. Thus they might well uncritically say that A and A' or A'' express the same thought, and that the same goes for the sentences S, $Q_{2^{\aleph_0}} x(x = x) \leftrightarrow Q_{\aleph_1} x(x = x)$ and CH. They might call such sentences 'essentially equivalent' or some such. But when our interest is in the precise demarcation between mathematics and logic, it is important to tread carefully. The sentence A is an arithmetical truth, because its formalization $f(a, b) = c$ is not valid on all models. The sentence A'' in contrast, or for that matter its first-order version A' obtained by dropping the two leading second-order quantifiers, is logically true. Similarly, CH is not a logical truth, since it is existentially committing; indeed, it turns out to not be provable in first-order ZFC, never mind first-order logic without non-logical axioms. In contrast, S is a logical truth if true. Responding to this or closely related objections, philosophers have been wont to concede that there is no boundary between logic and mathemat-ics (e.g. Shapiro 1998, p. 146). But—and this is another of this chapter's morals—that concession would be premature. According to isomorphism invariance, there *is* a principled boundary between logic and mathematics: sentences such as S lie on one side of it, whereas sentences such as CH lie on the other.

9.8 Identity

At this point, there is a possible objection that we must address. Our responses to the critics of isomorphism invariance have all involved relations between simple arithmetic claims and first-order logical truths. But these logical truths all involve numerically definite quantification and so rely crucially on the logical nature of identity.

Our reply is twofold. First, the logical status of identity is widely accepted and largely uncontentious. Most theoretical accounts of logical constanthood endorse it as logical, including not just invariantist but also inferentialist accounts.[24] Finally, the arguments against identity being logical are thin on the ground and unpersuasive. Nevertheless, we recognize that in this context our first reply is dialectically controversial. In particular, Feferman—one of the main proponents of the overgeneration objection—denies the logical nature of identity and puts forward a strong-homomorphism-invariance criterion which rules identity out as logical (as we saw in Chapter 8).

Our second reply is therefore that although our simple arithmetical examples have so far involved the logical nature of identity, other examples are available that do not rely on this assumption. Consider the following pair of statements: first G, viz.

$$3 \geq 2,$$

and a statement G' interpreting

$$\exists x \exists y \exists z (Fx \wedge \neg Fy \wedge Gy \wedge \neg Gz \wedge Hx \wedge \neg Hz) \rightarrow \exists x \exists y (Fx \wedge \neg Fy)$$

We may take G' to say that if there are at least three things that pairwise have different properties then there are at least two things that have different properties. So if the mathematical sentence G is true, then G' is logically true. The situation is analogous to CH and S but, crucially, G' does not rely on the logical nature of identity: G' only includes vocabulary that Feferman would accept as logical. So, if you are at all unsure about the logical nature of identity, then you can recast some of our earlier arguments in terms of G and G', rather than A and A'.

9.9 Conclusion

We began with the vague worry that logic may have been rendered *sensitive* to mathematics if we accept this demarcation. The worry has been shown to be illusory, however: there is no good sense of 'sensitive' relating logic to mathematics in a problematic way. We have worked on our opponents' behalf to spell out the notion of 'sensitivity' (or 'entanglement') in play. We used the example of the Continuum Hypothesis because this is the example usually brought up in the literature, but our discussion easily generalizes. *Mutatis mutandis*, our discussion in this chapter parries the objection that logical constants or extensions are

---

[24] For example, Read (2004) and Milne (2007).

sensitive in a problematic way to mathematical ontology. Many of the arguments we have put forward can also be used to defend second-order logic as logic, as our presentation makes clear. And we have also noted that the boundary between mathematics and logic is firmer than is often thought. We relegate to an appendix another, more technical, version of the overgeneration objection, and move on to the absoluteness objection (Chapter 10) and the intensional objection (Chapter 11).

Appendix: logical notions

In Chapter 9, we argued that no version of the overgeneration argument against isomorphism invariance is convincing. We end our discussion of overgeneration by considering a final version of this worry. As Bonnay (2008, pp. 36–7) sees it, the overgeneration argument is not limited to statements/sentences. The charge is that isomorphism invariance classifies certain *notions* that are in fact mathematical as logical. We have focused on the logical truth of various statements but, so this objection goes, that is of secondary concern: the real worry is that a *notion* such as 'there exist uncountably many' has been deemed logical, when intuitively it shouldn't.

We respond to the general version of this objection first, before considering a particular topological example offered by Bonnay. An important theme of this chapter has been the need to keep various closely related statements separate. It is tempting to say, for example, that S and CH express the same content, or similar. We argued in §9.7 that this sort of talk is incorrect, although understandable in the context of mathematics. When we avoid it, overgeneration worries dissolve. Our reply to the present objection is similar: just as it is important not to conflate various *statements*, it is important not to conflate various *notions*.

The objection that some notion ought or ought not to be classified as logical is, we believe, much harder to assess than the objection that some statement (such as CH) ought or ought not to be classified as logical. This is natural, since we have clearer (though still messy) intuitions about the extension of 'logical truth' than 'logical constant'. It would be very difficult, for example, to introduce a novice to the concept of logical constants before explaining the concept of logical truth and consequence. We believe that any plausible attempt to draw a principled line between logical and mathematical notions will have to rely at least in part on judgements about various statements' logicality. In as much as we do have a pre-theoretic grip on whether a *notion*, as opposed to a *statement*, is logical or not, it is not at all clear that we—meaning mathematicians and the mathematically well-educated—tend to think of general mathematical notions as non-logical. The notion of a topological space, or of a continuous function between two topological spaces, or of a group, or of a homomorphism between two groups, and so on, are taken to be maximally general notions whose instances are anything whatsoever that satisfies the defining conditions. Naturally, the study of such notions belongs to mathematics. But the notions themselves—as opposed to claims that something instantiates them—are entirely general and apply to anything that satisfies some structuring conditions, as David Hilbert was one of the first to note.

What is true is that one can perfectly well express CH, to within a reasonable informal standard of mathematical equivalence, by using various generalized quantifiers. We have suggested that this does not commit one to set-theoretic objects (e.g. infinite cardinals) because these quantifiers are isomorphism-invariant primitives. But are not ontological commitments concealed by such a move? Say we can only understand or justify assertions involving the quantifier 'there are continuum many' by relating its subject matter by a set isomorphism to a canonical example of continuum size. Doesn't it follow that the quantifier 'there are continuum many' carries mathematical commitments?

Our retort to this charge is threefold. First, as noted in §9.5, mathematics can be used to yield knowledge of logical truths; as mentioned, this applies to first- as well as second-order logical truths. And as also noted earlier, there is no relevant disanalogy between cardinality quantifiers expressible in first-order terms such as $\exists_{512}$ on the one hand and higher-order ones such as $Q_{>\aleph_0}$ ('there are uncountably many') on the other. Arithmetic is no less part of mathematics than set theory, the familiarity of first-order logic notwithstanding. Our contention is that both these quantifiers are logical, and we see no principled reason to regard the latter as 'mathematical' or 'problematically mathematical' but the former not. Second, we deny that understanding an assertion involves ontological commitment to the entities over which it quantifies. For example, we can perfectly well *understand* discourse about fictional objects (e.g. dragons or fairies), or metaphorical language, without thereby incurring commitment to the objects described.

Finally, the use of set theory to grasp a sentence such as S is only a crutch, although admittedly a very helpful one. The sentence S can in principle be understood *literally*, using only the conceptual resources expressed by a second-order language free of non-logical terms and without deploying any further concepts. In much the same way, the statement A' in §9.2 can be understood by deploying the conceptual resources of first-order logic, without ever entertaining the thought that $2 + 5 = 7$. We realize, of course, that it is very hard for limited human cognizers to understand S without using set theory as a crutch. It would be just as hard for us to understand the $10^{10^{10}}$-termed Boolean connective $*_{Fermat}$, which outputs the truth-value True iff the number of True inputs equals N for N the index in a true Fermat equation,[25] without recourse to number theory. But we take it that the philosophically interesting notion of conceptual dependence at stake here is an in-principle one that abstracts from contingencies of human implementation rather than an in-practice notion that depends on the specific mechanisms of human cognition.

We turn finally to Bonnay's example of a continuous function between topological spaces, intended to make his objection sharper. Briefly stated, the notion

$$(X, \tau_X) \text{ and } (Y, \tau_Y) \text{ are topological spaces and } f : (X, \tau_X) \to (Y, \tau_Y) \text{ is continuous}$$

may be seen to be isomorphism invariant when one unpacks the definitions (here τ_X and τ_Y are the topologies on X and Y respectively). The reason is that for the function $f : (X, \tau_X) \to (Y, \tau_Y)$ to be continuous, it matters not what the elements of X or Y are, but only that the pre-image under f of any element of τ_Y is an element of τ_X, a fact that is preserved under isomorphism.[26] More generally, any class of structures closed under isomorphism gives rise to a logical notion that applies to all and only elements of that class.

Two replies may be made to this challenge. First, it seems to be an inherent feature of the invariantist account. The invariantist carves up the space of models into equivalence classes, identifying logical notions as (all and only) those that respect the equivalence

---

[25] N is an index in such an equation iff there are positive integers x, y, z such that $x^N + y^N = z^N$. Andrew Wiles' proof shows that $*_{Fermat}$ outputs True iff exactly 0, 1, or 2 of the inputs are True.

[26] Meaning that if the topological spaces (X, τ_X) and $(X', \tau_{X'})$ are isomorphic—i.e. homeomorphic—via the bijection $i_{XX'} : X \to X'$, and (Y, τ_Y) and $(Y', \tau_{Y'})$ are homeomorphic via the bijection $i_{YY'} : Y \to Y'$, then $f : (X, \tau_X) \to (Y, \tau_Y)$ is continuous iff $f' : (X', \tau_{X'}) \to (Y', \tau_{Y'})$ is continuous, where $f' = i_{YY'} \circ f \circ i_{XX'}^{-1}$.

class structure (i.e. they apply in some way to a particular model iff they apply in the same way to all its equivalents, for the given notion of equivalence). To illustrate the point briefly, consider Bonnay's own proposal, sketched in §8.4, according to which the correct equivalence relation is not isomorphism but potential isomorphism (defined there). As mentioned in §8.4, this account makes the notion 'structure potentially isomorphic to $\langle \mathbb{R}, < \rangle$' logical; in this case, the class includes not just structures isomorphic to $\langle \mathbb{R}, < \rangle$ but also structures such as $\langle \mathbb{Q}, < \rangle$.

Second and relatedly, determining whether or not some function between topological spaces is continuous or not seems to be a matter of applying definitions. In this intuitive sense at least, it *is* a logical notion. Take, for example, the topological spaces X and Y and function f between them given by:

$X = \{a, b\}$ and $\tau_X = \{\varnothing, X, \{a\}, \{b\}\}$;
$Y = \{c, d\}$ and $\tau_Y = \{\varnothing, Y, \{c\}\}$;
$f(a) = c; f(b) = d$.

As a matter of first-order logic, the conclusion that f is continuous follows from these conditions. More precisely, consider the conditional whose antecedent is the conjunction of the above facts about (X, τ_X), (Y, τ_Y), and f in appropriately unpacked set-theoretic notation (so that for example $X = \{a, b\}$ becomes $\forall z(z \in X \leftrightarrow z = a \lor z = b)$), and whose consequent states that f is continuous, that is, the pre-image under f of every element of τ_Y is an element of τ_X. Availing ourselves of the usual quantifier restriction conventions, the conditional's consequent is the statement

$$\forall O_Y \in \tau_Y \exists O_X \in \tau_X (\forall x \in O_X(f(x) \in O_Y) \land \forall y \in O_Y \exists x \in O_X(y = f(x)))$$

The resulting conditional is a first-order validity. We may also formalize the statement of f's continuity in a higher-order logic.[27] Either way, no extra-logical, somehow purely mathematical, premise is needed to determine f's continuity (given f's existence).

In sum, if the type of facts required to classify any given function from one topological space to another as continuous are what determine whether the notion of a continuous function is logical or not, then this notion is indeed logical since nothing but logic is required for the classification. Contrast this with an empirical notion, say that of a car: to determine whether some specified object is a car one needs empirical information about it. Of course, in contrast to the simple examples (X, τ_X) and (Y, τ_Y) given above, some topological spaces are not first-order definable, so the notion of a continuous function between such spaces will not admit first-order characterization. But what is not first-order expressible in these cases is the description of the spaces themselves or the description of the function in question, not the fact that the function is continuous, which remains the (first-order) consequent of the conditional, as specified above. The same point generalizes to other mathematical structures and maps between them (such as, for example, groups and group homomorphisms, and so on).

---

[27] So that τ_X and τ_Y become second-order predicate constants. The consequent becomes $\forall F(\tau_Y F \rightarrow \exists G(\tau_X G \land \forall x(Gx \rightarrow Ff(x)) \land \forall y(Fy \rightarrow \exists x(y = f(x) \land Gx))))$.

10

The Absoluteness Objection

We've now defended isomorphism-invariance accounts of the logical constants against the heterogeneity charge (Chapter 8) and the overgeneration charge (Chapter 9). In the present chapter, we defend them against a third objection.

10.1 The absoluteness objection

As noted in Part II, our hypothesis that the one true logic has sort $\infty\infty$ is closely related, although not identical, to the Tarski–Sher thesis. And as noted in Chapter 8, Feferman is without doubt Tarski–Sher's most influential critic. We have responded to his objection that 'no natural explanation is given by the thesis of what constitutes the *same* logical operation over arbitrary domains' in Chapter 8, to the objection that the Tarski–Sher thesis 'assimilates logic to mathematics' in Chapter 9, so we now turn to his third objection, that the thesis involves notions that 'are not set-theoretically robust, that is, not absolute'.[1] As we shall see, this *absoluteness* objection connects interestingly with the realism/anti-realism debate about set theory.

First, let's recall some definitions from elementary set theory. If $\mathfrak{A}$ is a substructure of $\mathfrak{B}$ with the same signature and ϕ is a formula in this signature with n free variables, to say that ϕ is absolute for $\mathfrak{A}$, $\mathfrak{B}$ means that $\phi[\bar{a}]$ holds in $\mathfrak{A}$ iff $\phi[\bar{a}]$ holds in $\mathfrak{B}$, for any n-tuple $\bar{a}$ whose elements are all in the domain of $\mathfrak{A}$. We usually say that ϕ is absolute *simpliciter* when the signature is that of set theory, $\mathfrak{A}$ is a transitive set in which the interpretation of $\in$ is the membership relation, and '$\mathfrak{B}$ is the universe of sets and also interprets $\in$ as membership'. Of course, $(V, \in)$ is not a structure in standard set theory without classes, so this is shorthand for saying that ϕ is absolute just when $\phi[\bar{a}]$ holds in $\mathfrak{A}$ iff $\phi[\bar{a}]$ is true, with $\bar{a} \in A^n$ (A being the domain of $\mathfrak{A}$).

Although isomorphism of structures is defined in terms of the structures themselves and makes no reference to language (as opposed to, for example, elementary equivalence), the existence of an isomorphism between any two structures depends on the ambient universe of sets. Intuitively, $\mathfrak{A}$ and $\mathfrak{B}$ may not be isomorphic in some model that lacks an isomorphism between the two, even if

---

[1] All three quotations are from Feferman (2010a, p. 4). This 2010 article reprises Feferman (1999).

One True Logic: A Monist Manifesto. Owen Griffiths and A.C. Paseau, Oxford University Press.
© Owen Griffiths and A.C. Paseau 2022. DOI: 10.1093/oso/9780198829713.003.0010

these structures *are* isomorphic in a bigger model which includes the missing isomorphism. To illustrate this with a simple example, consider a structure $\mathfrak{A}$ with domain α, where α is some countably infinite ordinal not equal to ω, and a structure $\mathfrak{B}$ with domain the least infinite ordinal ω. In some countable transitive model M of ZFC, $\mathfrak{A}$ and $\mathfrak{B}$ may not be isomorphic, for example if M does not contain a bijection from α to ω (so that α is uncountable in M), though $\mathfrak{A}$ may be isomorphic to $\mathfrak{B}$ in some generic extension of M (that satisfies the axioms of ZFC). In other words, the relation of isomorphism between models is not absolute.

Feferman and Bonnay both object that the correct notion on which to base an invariantist criterion of logicality should be absolute (or 'robust' as Feferman sometimes calls it).

> The good [i.e. correct] notion of invariance for logicality is to be provided by a similarity relation S such that S is absolute with respect to ZFC.
>
> (Bonnay 2008, p. 60)

> The set-theoretical notions involved in explaining the semantics of $\mathcal{L}_{\infty\infty}$ are not robust.   (Feferman 2010a, p. 8).

Yet they offer no motivation other than the desire to avoid making sameness of structure dependent on the background universe of sets. It seems to us that there could be three possible motivations behind it (one suggested by Feferman, two undeclared), which we consider in turn.

10.2 Meaning-theoretic motivations

First, Feferman seems to hold that we should demand absoluteness for meaning-theoretic reasons. In the main place where he discusses the absoluteness objection, he writes that it

> is in a way subsidiary to [the overgeneration objection]. The notion of 'robustness' for set-theoretical concepts is vague, but the idea is that if logical notions are at all to be explicated set-theoretically, they should have the same meaning independent of the exact extent of the set-theoretical universe. For example, they should give equivalent results in the constructible sets and in forcing-generic extensions. Gödel's well-known concept of *absoluteness* provides a necessary criterion for such notions, and when applied to operations defined in $\mathcal{L}_{\infty\infty}$, considerably restricts those that meet this test. For example, the quantifier 'there exist uncountably many x' would not be logical according to this restriction, since the property of being uncountable is not absolute.
>
> (Feferman 2010a, p. 9; compare also Feferman 1999, p. 38.)

Feferman is thus explicit that this objection is *subsidiary* to the overgeneration problem, but does not explain what this means. An obvious interpretation is that the absoluteness objection is subsidiary in the sense that it is of secondary importance to the overgeneration argument: the absoluteness problem is a problem, in other words, because it has extensional upshots, namely overgeneration. This seems like the correct interpretation: Feferman admits that the absoluteness objection is 'vague' but the overgeneration argument is relatively clear, since precise cases can be given. Certainly, Feferman spends much more time on the overgeneration than the absoluteness objection. The absoluteness argument should, therefore, not worry us, if it is only *really* a problem because of its extensional upshot. We argued at length against the latter in Chapter 9, so we should not be worried by the former.

If this is not the correct interpretation of the relationship between absoluteness and overgeneration, we can still respond. The worry seems to be primarily about *meaning*: if logical constants are to be explicated in anything like invariantist terms, Feferman writes in the above quotation, 'they should have the same meaning independent of the extent of the set-theoretical universe'. Why should we think that isomorphism invariantism fails this condition? The example given is the quantifier 'there exist uncountably many' which, we are told, 'would not be logical according to this [absoluteness] restriction'. Feferman is correct that the *uncountably many* quantifier would not be declared logical by the absoluteness restriction, since his own invariantist criterion is designed to respect absoluteness and this quantifier fails his test. But he does nothing to connect this point about absoluteness to the previous point about meaning.

We do not believe that the meaning of the quantifier 'there exist uncountably many' is ambiguous on the isomorphism-invariance approach. In ordinary mathematics, where it is most often found, the quantifier has a clear sense or intension. It applies for example to the real numbers but not to the rational numbers, to the transcendental numbers but not to the algebraic numbers, to the functions from the natural numbers to $\{0, 1\}$ but not to the functions from $\{0, 1\}$ to the natural numbers, and so on. We may model the semantic value of this quantifier, of type $((i))$, by stating that it applies to a unary property P of type (i) iff P applies to uncountably many individuals in the domain. Similarly if we model the quantifier as an operation from n-tuples of sets of variable assignments to a set of variable assignments, or in some other way. We have been given no reason to believe that this semantic value will be different depending on the extent of the set-theoretic universe, so we have been given no reason to believe that the meaning of the operator will be different. It is in this way analogous to the meaning or intension of an empirical predicate, such as 'is red': although this predicate's extension may vary depending on which things are red, its intension does not so vary.

Perhaps the response will be that the semantic values of cardinality quantifiers are functions whose inputs and outputs are set-theoretic objects so their meaning is sensitive to what sets there are. If *this* is sufficient, however, for the meaning of an expression to be ambiguous, then the meanings of first-order logical constants are ambiguous. This is because the extensions of all logical constants can be treated in the same way, for examples as functions from n-tuples of sets of assignments to sets of assignments. If such set-theoretic extensions are problematic in the case of cardinality quantifiers, they should be problematic everywhere. But Feferman accepts this framework, and he accepts, for example, the first-order quantifiers as logical. All he rejects is the logical nature of expressions definable in $\mathcal{L}_{\infty\infty}$ but not $\mathcal{L}_{\omega\omega}$. We have so far not been given any reason to believe that we should demand absoluteness for meaning-theoretic reasons. Our use of 'there exist uncountably many' and related quantifiers and notions seems to be in good semantic order.

10.3 Anti-realist motivations

The second motivation might be anti-realism about set theory. To see why, observe that the absoluteness condition—the demand that the invariantist criterion of logicality be based on an absolute notion—seems to be incompatible with realism about set theory. For our purposes, let *set realism* be the thesis that there is a single universe of sets and that ZFC, as usually interpreted, is true of this universe.[2] In contemporary philosophy of set theory, such a view is often known as 'universism', or to be more precise, 'determinate universism' as it takes set-theoretic sentences to be determinately true or false. If set realism holds, then whether two structures are isomorphic or not is a question that has a yes or no answer, independent of any background model. For example, if α is any ordinal not equal to ω, the structure $\langle \alpha, < \rangle$ is not isomorphic to $\langle \omega, < \rangle$, because there is no order-preserving bijection from α to ω, a fact provable in ZFC. According to set realism, then, the isomorphism account returns a univocal answer to whether some operator is logical or not: either an isomorphism exists in the (single, uniquely intended) universe of sets, or it does not. To oppose this idea, one must be an anti-realist about set theory (i.e. oppose realism in the sense just defined).

Anti-realism about set theory, however, is problematic in several ways in the present context. First, it sits uncomfortably with the standard, model-theoretic conception of logical consequence, which the invariantist criterion of logicality seeks to complement. For the notion of logical consequence to be univocal, there

---

[2] No commitment is made to ZFC's completeness; quite the contrary, set realists believe that it is an incomplete theory of the universe. We could of course go into more detail about what 'ZFC' is—e.g. the first-order or second-order version?—but these clarifications are not needed for our main point here. We also note, incidentally, that our use of 'realism' differs from that in the universism vs. multiversism debate in the philosophy of set theory, in which both sides claim to be realists. We could have called our view 'determinate universism', as we do in the next sentence, but instead call it 'realism' for simplicity and because that fits better with the broader use of the term within philosophy more generally.

must be a univocal account of what models exist, that is, what sets or classes exist, since this is what we take models to be. One response to this might be to essay a model theory without set theory; another to adopt logical pluralism, of the type discussed in Part I for example. But if either of these responses, or another, is what underpins the critique of the isomorphism-invariance account of logicality, there is no hint of it in the literature under discussion.

Second, set realism is a well-established account of set theory. It is prevalent not just among set theorists but also among mathematicians, logicians, and philosophers of logic and mathematics. Feferman is well-known for having doubts about set realism (see e.g. Feferman 2000), although he does not cite them in either of his critical papers on isomorphism invariance (Feferman 1999, 2010a), so it is hard to tell whether his case ultimately rests on these doubts. Clearly, we cannot take on the large task of evaluating set realism here. We note merely that attempts to dispense with the assumption that there is, ultimately, a universe of sets, such as Hellman's (1989) eliminative structuralism or Field's nominalism (1980, 1989) are, at best, highly controversial. Similarly with more recent multiversist views, such as that of Hamkins (2012, 2015), which couple belief in the objective existence of set universes with the idea that there are many of them. If something like this is what the absoluteness argument relies on, that argument is much more problematic than has hitherto been realized. At the very least, it ought to be openly acknowledged as buttressing the objection.

Related to the last point, we note in passing that the notion of two structures being isomorphic, although not absolute, *is* absolute for initial segments of the hierarchy. More precisely, suppose $\mathfrak{A}$ and $\mathfrak{B}$ are two first-order structures and γ is an ordinal with $\mathfrak{A}, \mathfrak{B} \in R(\gamma)$, $R(\gamma)$ denoting the sets of rank less than γ in the hierarchy of well-founded sets. We write '$\cong$' for the isomorphism relation, so that $\mathfrak{A} \cong \mathfrak{B}$ means that there is an isomorphism between $\mathfrak{A}$ and $\mathfrak{B}$; and if we superscript this formula with $R(\gamma)$, thus: $(\mathfrak{A} \cong \mathfrak{B})^{R(\gamma)}$, we thereby indicate that this relation holds when all the quantifiers are relativized to $R(\gamma)$. Then $(\mathfrak{A} \cong \mathfrak{B})^{R(\gamma)}$ iff $(\mathfrak{A} \cong \mathfrak{B})$, from which we deduce that an isomorphism between $\mathfrak{A}$ and $\mathfrak{B}$ exists iff it exists in $R(\gamma)$.[3] The moral: even if one believes that there is not one true set theory, but that set theory is about all iterative models of ZFC, then whether or not two given sets are isomorphic is a determinate fact. It is not a fact that changes from one such model to another (unlike, say, the existence of various large cardinals).

The third and final point in connection with anti-realism is that the absoluteness constraint does not seem to be naturally motivated by set anti-realism either. If ZFC is not a correct account of facts about models (sets), why should the correct invariantist notion be ZFC-absolute? Bonnay appears to recognize that there is a problem here, but does little to dispel it other than to point out that

---

[3] Or as it is sometimes put, $R(\gamma) \preceq_{\cong} V$ (V here is the class of all sets). See e.g. Kunen (2013, pp. 97–8) for more details.

ZFC is the 'standard background set theory' (2008, p. 60). ZFC is indeed the standard background set theory, but for the anti-realist that is presumably a sociological fact. We do not see how to convert this sociological observation into a philosophical justification without some story about how ZFC manages to latch onto, or at least correlate with, mathematical reality. Feferman considers various versions of Kripke–Platek (KP) set theory, also mentioned by Bonnay (2008, p. 60). From a philosophical perspective, however, it is hard to see what the significance of KP might be. A thoroughgoing anti-realist cannot privilege *any* set theory as a guide to what sets exist since, by assumption, there are either several equally good set theories or no facts about what sets exist. By their own lights, the critics should not merely 'minimize the dependency of logic on set theory' (Bonnay 2008, p. 60), but they should reduce it to *nil*.[4] Perhaps the idea is instead that KP is to be trusted about sets but that ZFC is not? This timid form of set realism is hard to square with present mathematical practice. Feferman (2010a, p. 17) somehow thinks that there is a problem with infinite sets, since he is keen to assume no special set-theoretical assumptions beyond those needed to generate the hereditarily finite sets. But again, we find this a halfway house: if the assumption of sets in the metatheory is problematic, then no sets whatsoever should be assumed, and presumably invariantism must be abandoned; or if it is unproblematic, it is hard to see what sustains the absoluteness requirement.

A final observation.[5] From the infinitary logician's perspective, the nonabsoluteness of notions such as 'there are infinitely many' is a product of the extraordinarily weak—finitary—background logic. Once we move to a logic of sort $\infty\infty$ extending $\mathcal{L}_{\infty\infty}$ (i.e. FOL$_{\infty\infty}$), as we advocated in Part II, all cardinality quantifiers are definable and thus absolute. Indeed, if set theory is recast in a logic allowing κ-ary conjunction and disjunction for every κ (i.e. of sort at least $\infty\omega$), then each rank of the iterative hierarchy is also definable.[6] As long as the logical notions are respected, there is then no such thing as non-standard models of the sentences expressing the notions Feferman and others find objectionably

---

[4] Bonnay informally distinguishes between 'problematic' and 'unproblematic' set-theoretic content (2008, p. 56 and elsewhere), but neither defines nor explains the distinction much. At times, it seems that problematic notions are precisely those that are not absolute: '...*Iso* is a typical example of a set-theoretically problematic similarity relation, whose extension depends on the specific features of the model of set theory one is working with' (2008, p. 56). At others, it seems that problematic notions are those that do not flow from our semantic competence (2008, p. 65). We are suspicious in general of the idea that pure semantic competence underwrites inferential abilities: see Williamson (2007, pp. 73–133 and 281–5) for a sustained critique. We are suspicious in particular of Bonnay's apparent belief that they underpin our basic inferential abilities in arithmetic and in handling the finite/infinite distinction.

[5] Compare Sher (2013, p. 180). See also Chapter 10 of Sher (2016) for a discussion of the absoluteness objection complementary to our own.

[6] Define $V_0(x)$ as $x \neq x$ and $V_\alpha(x)$ as $\forall y(y \in x \rightarrow \bigvee_{\xi<\alpha} V_\xi(y))$. The hierarchy up to α may then be defined by the following sentence:

$$\forall x \forall y[x = y \leftrightarrow \forall z(z \in x \leftrightarrow z \in y)] \wedge \forall x \bigvee_{\beta<\alpha} V_\beta(x)$$

nonabsolute. Of course, this observation does not provide dialectical leverage against these philosophers, since it depends on having made the transition from a first-order theory of models to one cast in an infinitary logic. Anyone sufficiently troubled by the absoluteness objection will decline to make this transition. But it does show that, from the perspective of the infinitary logician who *has* made the transition, the absoluteness objection can be diagnosed as a symptom of relying on an overly weak logic. For the infinitary logician, the move to a logic of sort $\infty\infty$ undermines the absoluteness objection. It can't even be formulated against her.

10.4 Independence motivations

Let us look in another direction. A final motivation behind the absoluteness condition might be that the analysis of logicality should be independent of mathematics, or set theory in particular. Feferman (2010a, p. 4), for instance, explicitly asserts that to avoid the absoluteness objection, 'one should restrict to definitions that are absolute with respect to a system of set theory that makes no assumptions about the size of the universe'.

This third potential motivation is just as problematic as the previous two. To begin with, it sits equally uncomfortably with the model-theoretic conception of logical consequence, since our best theory of models, namely set theory, is an unabashedly mathematical theory. Second, the constraint, if applied consistently, would presumably prohibit any analysis of logicality that deploys resources going beyond logic. It is hard to see why various versions of Kripke–Platek set theory, weaker than ZFC though they be, are an improvement in this respect.[7] Third, the motivation is much more naturally married with inferentialism than invariantism.[8]

A broader, final, point is that this insistence on what might be called purity of analysis seems to us misguided and a barrier to progress. Roughly, a 'pure' analysis of notions drawn from a particular D assumes nothing other than notions available in that domain; for example, if geometry is the domain, a pure analysis would be one that employs only geometric notions.[9] Purity of analysis or proof is an idea that has emerged from reflection on mathematics. Mathematics, however, is replete with examples of ideas and techniques from one subfield being used to elucidate another, and is much the better for it. There is no immediate reason why

---

[7] It could be argued that some weak theories are equivalent to very weak logical systems that barely transcend syntax, as suggested by Feferman (2010a, p. 17). But the reply would surely be that there is no robust sense in which set-theoretic assumptions are special and others generic. Once again Feferman's idiosyncratic views about set theory discussed in §9.4 may be what underlie the discussion here. If so, they should be brought to the surface.

[8] Feferman hints at this at the end of his (2010a); see pp. 17–8 therein.

[9] Detlefsen (2008) is an introduction to purity/intrinsicness as an ideal of proof.

the analysis of logic should be any different. It certainly does not follow from the fact that logicality is analysed set-theoretically that its epistemology is that of set theory. As Frege was well aware, the fact that Euclidean geometry can be reduced to Cartesian or analytic geometry does not imply that the epistemology of geometry is that of analysis.

10.5 Conclusion

We have now considered the objection that the set-theoretic notions required to express our invariantist position are not absolute, which the objectors equate with not being robust. We accepted that the notions are not absolute, but denied that they ought to be. We considered three different motivations for this demand, and found none of them convincing. One problem remains, the subject of the next chapter.

11

The Intensional Objection

In this chapter, we consider a well-known charge against taking isomorphism invariance to be a *necessary* and *sufficient* account of logical constanthood. The charge is that this account fails to *intensionally* capture the logical constants. Concepts pick out their extension via their intension. Intensional objections to isomorphism invariance claim that the intensions of logical constants are ignored. Generally, they begin by noting that isomorphism invariance renders logical constanthood an entirely extensional matter, ignoring meaning. But there are many ways of picking out an isomorphism-invariant extension, some of which look logical and some of which don't. The challenge to the invariantist is to rule out these non-logical-seeming ways of picking out an invariant extension.

Fortunately, we are not committed to the account of logical constants the challenge targets. All we committed ourselves to in Chapter 6 was that for any isomorphism-invariant *relation* (a worldly entity) there must be a logical *constant* (a linguistic one) that denotes it. This, and no more, was required for Part II's 'top-down' argument. Thus we are not committed to isomorphism invariance being sufficient for logical constanthood; for all we have said, an expression may be extensionally equivalent to a logical constant but still not be a logical constant. The identity relation, for instance, is isomorphism invariant and so must be denoted by a logical constant; but not every expression that denotes the identity relation need be a logical constant. Nor are we strictly speaking committed to isomorphism invariance being necessary for logical constanthood; for all we have said, there may be expressions that have isomorphism-*variant* extensions and yet are still logical. The one true logic may abound with even more logical constants than dreamt of in Part II.

Our main work in this book is therefore done. The last of the principal objections to isomorphism-invariant approaches to logic, the intensional objection, targets the sufficient condition for logical constanthood. It does not affect us. Nevertheless, some philosophers seem to think that the objection calls into question the whole invariantist approach to logic and the logical constants. We'll therefore close the book by discussing it. In this spirit, we define an 'invariantist' in this chapter as someone who takes an expression to be logical iff its extension is isomorphism invariant.

One True Logic: A Monist Manifesto. Owen Griffiths and A.C. Paseau, Oxford University Press.
© Owen Griffiths and A.C. Paseau 2022. DOI: 10.1093/oso/9780198829713.003.0011

11.1 Unicorn negation

To the best of our knowledge, intensional objections to isomorphism invariance were first raised by McCarthy (1981). He argues that, on the isomorphism-invariance account, the 'logical status of an expression is not settled by the functions it introduces, independently of how those functions are *specified*' (1981, p. 516). McGee (1996) and William Hanson (1997) put forward similar objections to isomorphism invariance.

Like the majority of the subsequent literature, we'll focus on McGee's (1996) example of *unicorn negation*:

$$\mathcal{U}\phi =_{Def} (\text{not-}\phi \text{ and there are no unicorns})$$

Here ϕ is a variable instantiable by meaningful sentences. The word 'not' abbreviates 'It is not the case that', which by stipulation is the natural-language sentential operator corresponding to Boolean negation. We'll simply call it 'negation' for short. Since there are no unicorns, unicorn negation is coextensive with negation. If isomorphism invariance is sufficient for logical constanthood, then unicorn negation is logical because negation is. So '$\mathcal{U}$' is a logical constant. Why might this be worrying for the invariantist?

First, there is the rough worry that '$\mathcal{U}$' is in some sense *about* unicorns. When we explain the meaning of '$\mathcal{U}$', we have to say something about unicorns and nothing like this seems to be true for any of the standard logical constants. The worry then is that the meanings of logical constants shouldn't involve such entities. But, as we argued in Chapter 9, intuitions about the logical constants are unclear. Instead, we should focus on the cases of logical consequence and logical truth, which are at least sharper.

Unfortunately, accepting unicorn negation as a logical concept throws up problems for logical truth too. Because unicorn negation and regular negation are coextensive, any instance of the following (in which ϕ is replaced by an interpreted sentence) is logically true:

$$\mathcal{U}\phi \text{ iff not-}\phi$$

But now substitute an absurdity for ϕ, such as, say, 'Everything is non-self-identical'. Then the right-hand side is true in every interpretation, so the left-hand side is true in every interpretation. By definition, the left-hand side would be 'It's not the case that everything is non-self-identical and there are no unicorns', which implies 'there are no unicorns'. A logical truth shouldn't imply the non-existence of unicorns, so something has gone wrong.

In §11.2, we consider some modal responses to the problem and explain why they are unsuccessful. We'll then move on to consider a more successful response, from Gil Sagi, in §11.3. Finally, in §11.4, we consider an open question: what account of logical constants the invariantist should offer.

11.2 Modal responses

One obvious response on behalf of the invariantist is to insist that, to be logical, an expression's extension must be isomorphism invariant on every *possible* domain. Equivalently, a putative logical constant must meet the requirement of isomorphism invariance of *necessity*. This solves the problem because, while there are no unicorns in the domain of actually existing things, unicorns are metaphysically possible.[1] In a domain containing unicorns and non-unicorns, we will be able to map some unicorn to some non-unicorn, disturbing the extension of 'is a unicorn'. Negation and unicorn negation would not be coextensive on this picture, so the above argument doesn't get off the ground.

But the response won't do. McGee anticipates it and puts forward the following counterexample:

$$\mathcal{H}\phi =_{Def} (\text{not-}\phi \text{ and water is } H_2O)$$

If, like Kripke, we think that water is necessarily H_2O, then '$\mathcal{H}$' is likewise coextensive with '$\neg$' and so is logical for the invariantist. And yet, by exactly parallel reasoning with the unicorn case, it shouldn't be logical, since then any instance of '$\mathcal{H}\phi$ iff not-ϕ' is a logical truth and it shouldn't be logical because it entails 'water is H_2O', again by parallel reasoning. If you don't buy Kripke's example, use your favourite necessary but intuitively non-logical truth.

In a later paper, McCarthy (1987) replies on behalf of isomorphism invariance that the problem can be solved by invoking *epistemic* necessity. Even if water is necessarily H_2O, it isn't as a matter of epistemic (*a priori*) necessity. However, as Mario Gómez-Torrente (2002, p. 21) suggests, we can easily cook up similar counterexamples. Consider, for example, *vixen* negation:

$$\mathcal{V}\phi =_{Def} (\text{not-}\phi \text{ and there are no female vixens})$$

Plausibly, there are no epistemically possible worlds in which we find a male vixen, so '$\mathcal{V}$' passes the test for isomorphism invariance.

McCarthy (1987) anticipates this and his response is to replace epistemic with *logical* necessity. Logical necessity, however, looks equally unattractive. First, it's not clear that we can grasp a concept of logical necessity that is independent

---

[1] *Pace* Kripke (1980, pp. 24, 156–8).

of logical consequence. A standard understanding is that a sentence is logically necessary just if it is true by logic alone (a logical truth). But, on most accounts, logical truth crucially involves an account of the logical constants, which is what isomorphism invariance seeks to provide. So the invariantist can't adopt this standard understanding without their definition of logical truth being circular.

We could instead accept logical necessity as a primitive notion. Again, the invariantist should be loath to do this. First, it increases the ideology of our account of consequence and the logical constants. Second, the invariantist may seek to account for logical necessity in terms of formality. For example, if the invariantist completes their project of accounting for logical consequence entirely in terms of formality, then they can go from here to *define* logical necessity: a sentence is logically necessary just if logically true. This would achieve a reduction of modal talk to formal talk. Such ambitions are dashed if logical necessity is invoked as a primitive. Again, this restriction to the scope of the project may prove inevitable but, again, it is worth trying to do better.

Finally, and this is McGee's response, we could appeal to *semantic* necessity. With a semantic constraint placed on isomorphism invariance, McGee conjectures that:

> A connective is a logical connective if and only if it *follows from the meaning* of the connective that it is invariant under arbitrary bijections.
>
> (McGee 1996, p. 578, italics ours)

The notion of an extension's invariance *following from* an expression's meaning is obscure (as McGee admits, hence this being a conjecture). Indeed, if this is intended to mean *logically* follows from, then we're back to the circularity worry above. Further, as John MacFarlane (2015) notes, this constraint doesn't obviously rule out cases like '$\mathcal{V}$', expressing vixen negation, since 'no vixen is male' is presumably as much a semantic necessity as an epistemic one. There is, however, something right about McGee's idea, which we'll return to a little later.

Overall, then, the project of placing further *modal* constraints on isomorphism invariance in response to this problem seems problematic. There are other conceptions of necessity that we haven't considered, of course, but ultimately it seems that there will be counterexamples to anything but *logical* necessity, and logical necessity is unacceptable for the reasons discussed. The situation is a familiar one. In Quine's (1951) famous arguments against analyticity, he considers whether some modal notion can succeed in helping us define synonymy. He finds that no notion of necessity can succeed other than *semantic*, and that's circular. Analogously: no notion of necessity can succeed here but *logical* necessity, and that's the notion we're after. So let's explore alternative responses.

11.3 Biting the bullet

Perhaps the most interesting response to the problem of unicorn negation in the recent literature is due to Gil Sagi (2015). Her overall strategy is to bite the bullet: accept that any instance of '$\mathcal{U}\phi$ iff not-ϕ' is a logical truth, as the invariantist believes, and explain away the reasons for doubting its logical truth. She finds two reasons in McCarthy's writings to think that any instance of '$\mathcal{U}\phi$ iff not-ϕ' is not a logical truth. First, logical truths should be *necessary*. But any instance of '$\mathcal{U}\phi$ iff not-ϕ' could have been false, so any such instance is not a logical truth (she calls this the modal argument). Second, logical truths should be *a priori* knowable. But any instance of '$\mathcal{U}$ iff not-ϕ' is knowable only *a posteriori*, so no such instance is a logical truth (the epistemic argument).

Let's focus on the modal argument, and take the relevant instance of '$\mathcal{U}\phi$ iff not-ϕ' to be '$\mathcal{U}A$ iff not-A' (think of a particular A, if you like 'The cat sat on the mat'). Why think that '$\mathcal{U}A$ iff not-A' could have been false? At this point, we find Sagi's argument rather compressed so we'll take some time to spell out our version of what's going on. For ease of presentation, let's use the following abbreviation:

U   There are no unicorns.

Our presentation of the argument deviates significantly from Sagi's but we take its core to be essentially the same. The argument runs as follows:

1. 'U' is true at our world, w_1, accessible to itself—premise
2. 'U' is false at some world, w_2, accessible from w_1—premise
3. Logical truths are necessarily true—premise
4. Invariantists judge '$\mathcal{U}A$ iff not-A' to be a logical truth—premise
5. '$\mathcal{U}$' is coextensive with 'not' at w_1—1, def. of $\mathcal{U}$
6. '$\mathcal{U}A$ iff not-A' is true at w_1—5
7. '$\mathcal{U}$' is not coextensive with 'not' at w_2—2
8. '$\mathcal{U}A$ iff not-A' is false at w_2—7
9. '$\mathcal{U}A$ iff not-A' is a contingent truth at w_1—6, 8
10. '$\mathcal{U}A$ iff not-A' is not a logical truth—3, 9
11. The invariantist's criterion of logicality is false—4, 10

Sagi argues that this argument is invalid, since it relies on an equivocation over '$\mathcal{U}A$ iff not-A'. When we say that '$\mathcal{U}A$ iff not-A' is true at world w_1 but false at w_2, we are equivocating because '$\mathcal{U}A$ iff not-A' does not *mean* the same in each case. Because the contingent sentence 'U' is part of the meaning of '$\mathcal{U}$', sentences involving '$\mathcal{U}$' have different meanings at worlds where 'U' is true, like ours, and worlds where it isn't, like w_2. Sagi writes:

> Logical truths should presumably be true in all possible worlds, but only so long as their meanings are fixed. As vexed an issue as meaning is, that much is assured when it is assumed that logical truths are necessary truths. That is, talk of necessary truths that employs possible worlds assumes that sentences can be evaluated in possible worlds *given the meanings that they have.*   (Sagi 2015, p. 9)

In other words, when we assess the truth-value of a sentence at a world, we don't test sentences alone but rather pairs of sentences and interpretations. In the case of unicorn negation, we find that:

⟨ '$\mathcal{U}A$ iff not-A', $\mathcal{J}_1$⟩ is true at w_1, where 'U' is true.

⟨ '$\mathcal{U}A$ iff not-A', $\mathcal{J}_2$⟩ is false at w_2, where 'U' is false.

Crucially, $\mathcal{J}_1 \neq \mathcal{J}_2$, since '$U$' is true at w_1 and false at w_2, and the truth-value of 'U' plays a role in the interpretation of '$\mathcal{U}$'. This is sufficient to dissolve the problem, since we've been given no reason to believe that '$\mathcal{U}A$ iff not-A' is contingently true at our world, w_1: for that, it would have to take different truth-values at different accessible worlds *on the same interpretation*, and all we have here are different sentence-interpretation pairs taking different truth-values. Relating this to the above argument, line 9 does not follow from lines 6 and 8. All we can conclude from lines 6 and 8 is that '$\mathcal{U}A$ iff not-A' *on one interpretation* is true at w_1 and that '$\mathcal{U}A$ iff not-A' *on a different interpretation* is false at w_2, accessible from w_1. These do not jointly entail that '$\mathcal{U}A$ iff not-A' is contingent at w_1.

So McCarthy-style arguments fail to show that '$\mathcal{U}A$ iff not-A' is merely contingently true. McCarthy has given us no reason to suppose that there is a sentence which is logically true, according to the invariantist, but not necessarily true.

McCarthy runs another, *epistemic* argument against the invariantist. The conclusion here is that there is a logical truth, according to invariantism, which is not *a priori* knowable. On the assumption that logical truths are *a priori* knowable, this is a problem. Of course, on a liberal conception of logical truths such as ours (see Part II), whether logical truths are *a priori* knowable is a substantial commitment. It is more plausible to suppose that, say, first-order logical truths are all *a priori* knowable than that logical truths of a highly infinitary logic such as $\mathcal{L}_{\infty\infty}$ are *a priori* knowable. That said, the kinds of alleged counterexamples to the *a priori* knowability of logical truths we are concerned with here turn not on the one true logic's infinitary features but on the more mundane fact that unicorns don't exist. Just as well then that Sagi offers another similar argument in response to McCarthy here using *epistemically* rather than *metaphysically* possible worlds.

Let's take our *a posteriori* truth to be 'The cat sat on the mat' and label it A. Sagi invites us to think of *a priority* as truth in all epistemically possible worlds. Let A be true at epistemically possible world w_1 but false at epistemically possible

w_2, accessible from w_1. McCarthy's argument will then run analogously with our argument above, with A in place of U and the modal language interpreted as *epistemic* rather than metaphysical. Just as with the above argument, it will reach the conclusion that a sentence is an *a posteriori* truth at w_1 and so not a logical truth.

The solution is also analogous. McCarthy wants us to hold that a particular sentence is true at one epistemically possible world but false at another, hence *a posteriori*. But, as argued in the metaphysical case, this sentence must be interpreted in order to be truth-apt, and its interpretation varies across worlds. This is because an *a posteriori* truth has been included in its meaning. So all we have is a sentence that *on one interpretation* is true at an epistemically possible world and *on a different interpretation* is false at an epistemically possible world. And these do not entail that the sentence is *a posteriori* at any single world. In other words, the epistemic version of line 9 does not follow from the epistemic versions of lines 6 and 8, since the interpretation of the sentence in question varies.

We believe that these arguments against McCarthy, which we take to be expanded versions of Sagi's arguments, are both successful. Unfortunately, we cannot leave the debate here. We have argued that logical truth is an essentially *formal* notion (understood as topic-neutral), so the modal status of '$\mathcal{U}A$ iff not-A' is not our primary concern (though we accept that logical truth is necessary). What would go squarely against our view would be if '$\mathcal{U}A$ iff not-A' were logical but not topic-neutral.

So is unicorn negation topic-neutral? It depends how we understand topic-neutrality. Recall from §6.2.3 that there are two main thoughts in the literature about topic-neutrality:

Discrimination   A concept is topic-neutral iff it is not able to discriminate between different individuals.

Universality   A concept is topic-neutral iff it is applicable to reasoning about any domain whatsoever.

The relevant question for us to ask here is whether the concept *unicorn negation* is topic-neutral on either or both of the understandings given. First, is unicorn negation topic-neutral in the sense of Discrimination? Yes: unicorn negation does not discriminate between particular objects. Its extension is of course sensitive to the truth-value of 'there are no unicorns' but, with the truth-value of that metalanguage sentence fixed, Sagi has convinced us that it does not discriminate. When its sense is fixed, it is coextensive with negation and does not discriminate for the same reason. Similarly, unicorn negation is topic-neutral by Universality. Again, once the sense of '$\mathcal{U}$' has been fixed, it can be used, just like negation, to reason about any domain whatsoever.

So it seems that unicorn negation, and so '$\mathcal{U}A$ iff not-A', express topic-neutral concepts, by both tests for topic-neutrality. Nevertheless, we may worry that too much work is being done here by the process of 'fixing the sense' of unicorn negation. After all, 'fixing the sense' of unicorn negation crucially involves the truth-value of the metaphysically contingent sentence 'there are unicorns'. In this world, yes, unicorn negation is coextensive with negation, but in other worlds it is not. This, it may be argued, reveals a sensitivity to what exists that should worry us. If our concern is Discrimination, then perhaps there is a trans-world sense in which unicorn negation is discriminatory.

Here it is worth pausing to consider which test fits better with the isomorphism-invariance approach presented in Part II. The question is readily answered: the first, Discrimination, test for logicality. Recall that this is a test primarily for what MacFarlane calls 2-formality:

> **2-formal** logical notions and laws are indifferent to the particular identities of objects. 2-formal notions and laws treat each object the same (whether it is a cow, a peach, a shadow or a number).   (MacFarlane 2000, p. 64)

2-formality is clearly an expression of the topic-neutrality of logic in the sense of Discrimination. Some logical concepts are obviously topic-neutral in the sense of Discrimination: identity, universality, and negation. Some concepts obviously are not: redness, circularity, and elephanthood. Isomorphism invariance respects all of these clear-cut cases.

Other cases are much less obvious, like unicorn negation. We gave brief arguments that unicorn negation *is* topic-neutral in the sense of Discrimination, but worried that these may be inconclusive. We believe that unicorn negation is a peculiar, artificially designed concept, which it is difficult to have intuitions about. In these cases, we suggest that isomorphism invariance should be taken as our guide. It is an explication of 2-formality, which is topic-neutrality in the sense of Discrimination and part of its job is to decide borderline cases. It respects the intuitively clear cases and should be followed in these more challenging cases.

It could be objected that, yes, intuitions about unicorn negation are unclear, but it is uncontroversial that the concept *unicorn* is not topic-neutral, in any sense. And yet, as McGee argues, logical truths like '$\mathcal{U}A$ iff not-A' entail 'there are no unicorns'. A logical truth should not have plainly non-topic-neutral sentences amongst its consequences. Surely this gives us reason enough to reject '$\mathcal{U}A$ iff not-A' as topic-neutral, regardless of our *direct* intuitions about its status.

We agree that the concept *unicorn* is not topic-neutral, in any reasonable sense. 'Unicorn' is not, then, a logical expression and so neither asserting nor denying the existence of unicorns should be true *by logic alone*. So 'there are no unicorns' should not be a logical truth. We also agree that it is a problem if a sentence that we judge to be a logical truth entails 'there are no unicorns'. But there is no reason to think that '$\mathcal{U}A$ iff not-A' implies 'there are no unicorns'. That an expression (such as $\mathcal{U}$) has its sense fixed by a sentence (such as 'there are no unicorns') does not create any sort of logical entanglement. It is a familiar Kripkean lesson that we cannot in general make the jump from (i) the fact that the reference of an expression e_1 is fixed by means of some other linguistic expression e_2 to (ii) the alleged fact that the meaning of e_1 somehow involves that of e_2 (Kripke 1980, pp. 53, 57–60). Kripke focused on the case of names, whose reference is usually fixed by the reference of an associated description; he pointed out that in English, names are (almost always) rigid designators whereas their usual reference-fixing descriptions are not. The same moral applies here: unicorn negation has its sense fixed by the sentence 'there are no unicorns', but is not logically entangled with it.

We conclude that the alleged problems raised by unicorn negation do not undermine invariantism.

11.4 An open question

The major argument for the non-logical nature of unicorn negation—that it implies 'there are no unicorns'—fails. Yet the problem of predicates such as 'is a unicorn' remains for the invariantist. Isomorphism invariantism about logical expressions is the thesis that an expression is a logical constant iff its extension is isomorphism invariant. An unintuitive upshot of this is that any expression coextensive with a logical constant is also a logical constant.

How about the predicate 'is a unicorn' then? On the domain of things in this world, this is coextensive with non-self-identity. One response on behalf of the invariantist to the intensional objection is to dig their heels in and insist that only extension matters. Anything coextensive with a logical expression, the response would go, is logical. But this is an uncomfortable commitment, since 'is a unicorn' is not topic-neutral and is overwhelmingly not thought a logical constant. It would be better to have a story as to why 'is a unicorn' is not a logical constant. The invariantist thus needs to provide further conditions which, along with invariance, are sufficient. We will close this chapter by briefly describing two options.

One approach is to try to make use of the notion of *formalization*. Invariantists might slightly shift their position and urge that instead of looking at whether an expression's extension is isomorphism invariant, we must look at whether its formalization's interpretations are isomorphism invariant. (We abbreviate the latter by saying that the formalization itself is invariant.) They would then need an argument that 'is a unicorn' does not receive an invariant formalization whereas 'being non-self-identical', which is extensionally equivalent to it, does; or, if possible domains are included, that 'is a regular heptahedron' does not receive an invariant formalization whereas 'being-non-self-identical' does. Invariantists would also need an account of formalization, which we've only touched on in this book. One thought is to largely let grammar and meaning be our guides, the invariantist's hope being that they can lead to the formalization of 'is a unicorn' as U, a predicate whose interpretations vary. This would be in contrast to, say, 'being equal to', formalized as '=', a predicate with a fixed interpretation.

The worry with this sort of approach is twofold. First, any grammatical or semantical parsing invariantists might rely on is likely to be distinctly *philosophical*. The grammatical categories invoked in formalizing sentences into logic were introduced by philosophers and logicians. The grammatical test for formalization is therefore highly unlikely to be an independent one. This point is emphasized and evidenced in detail in Oliver (1999). As Oliver observes, even categories as fundamental to philosophers and logicians as 'subject' and 'predicate' seem to have been introduced—at least in the way we use them—by philosophers and logicians with a particular logical axe to grind. The lesson is that we must tread very carefully when drawing philosophical conclusions from apparently grammatical observations, since they often do not form an independent test.

Second, this sort of account is much better suited to inferentialist approaches to the logical constants than to our preferred model-theoretic one. An approach to logic motivated by broadly invariantist considerations, we argued in Part II, inevitably leads to a highly infinitary logic. Yet it is hard to imagine constraints on the formalization of infinitary languages emerging from syntactical studies of natural language. Moreover, even when we look at the vocabulary of first-order logic there is a worry. If we formalize 'x is identical to y' as '$x = y$', which has a fixed interpretation, then 'is identical to' will be considered logical, since its formalization only has isomorphism-invariant interpretations. If, however, we formalize 'is identical to' as, say, 'Ixy', then 'is identical to' will not come out as logical since its formalization will plainly have isomorphism-*variant* interpretations. A version of this problem even affects the major alternative to semantic, invariantist approaches to the logical constants, namely inferentialism. It is far from clear that the inference rules for the identity predicate can be stated in a form that meets most inferentialists' standards for logical nature. Stephen Read (2004, 2016) has tried to provide such rules for identity but see Griffiths (2014) and Griffiths and Ahmed (2021) for responses.

Summarizing both worries, it seems unlikely that *independent* grammatical or semantic criteria can be found to demarcate logical from non-logical terms. In the absence of such, we are left with an entirely circular directive: among the expressions with isomorphism-invariant extensions, formalize as logical terms those whose intension is logical and formalize the rest as non-logical. So we are back to square one.

Perhaps the most promising idea—and this is the second approach—is that of Gila Sher. Her full account is as follows:

> A constant is logical iff (i) it denotes a formal operator [one that is invariant under all isomorphisms of its argument-structures], and (ii) it satisfies additional conditions that ensure its proper functioning in a given logical system: i.e., it is a rigid designator, its meaning is exhausted by its extensional denotation, it is semantically-fixed (its denotation is determined outside rather than inside models and is built into the apparatus of models), it is defined over all models, etc. (Sher 2013, p. 176; this passage is italicized in the original).

Now the constraint that logical constants be 'rigid designators' (also found in Sher's earlier work, e.g. 1991, p. 64) is not, in truth, that illuminating. A predicate such as 'is a regular heptahedron' has the same modal profile as 'is non-self-identical' since neither of them denotes anything in any (metaphysically) possible world. Moreover, it is easy enough to rigidify descriptions whose intensions the invariantist would not wish to classify as logical; consider for example 'is an actual-world unicorn'. Sher takes the necessity in question to be formal necessity, which obviates this problem. But for her, formal necessity is a primitive notion, and we are keen to avoid this. That leads to a broader methodological point: the thought that logical consequence is modal, which as we saw in Part I has much support, is one that we accept but do not make the cornerstone of our account. Our cornerstone, or preferred foundation if you like, is a *formal* understanding of logical consequence. This motivates isomorphism invariance well, has at least as much historical support and has modal consequences. So, ideally, it would be better to avoid primitive modal notions in rejecting 'is a unicorn' as logical.

The part of Sher's condition (ii) we find most promising is that a logical expression's meaning is exhausted by its extensional denotation. This idea, we take it, is that 'is a unicorn' does more than denote the empty relation, whereas 'is non-self-identical' does not. Incidentally, the 'extension exhausts intension' clause seems to capture what is right about McGee's thought, discussed in §11.2. Paraphrasing somewhat, McGee's thought was that the meaning of a logical constant somehow entails that its denotation is isomorphism-invariant. The idea of entailment is misplaced, as we saw; the thought that a logical constant's intension is intimately related to its extension is better captured by Sher's formulation.

The 'extension exhausts intension' idea may be promising, but, patently, it requires further development. It would be hard to maintain, for example, that 'is non-self-identical' *directly* denotes its extension whereas 'is a unicorn' does not. Both seem to have a sense, and the question is whether a correct theoretical account of sense can distinguish the former's sense from the latter's in such a way as to underwrite their different logical status. Clearly, a good deal of work in the philosophy of language must be done if this approach is to fly. We conclude then that Sher's idea points to a promising way forward, but that there is still some way to go to fulfil its promise.

11.5 Conclusion

The arguments in this book relied on there being, for each logical relation, at least one logical constant denoting it. They did not rely on any other criteria of logical constanthood, intensional or otherwise. Nevertheless, we have considered intensional objections to isomorphism invariance that have appeared in the literature. The best-known problem is that of unicorn negation. Our response is to bite the bullet. Unicorn negation *is* a logical operation. This shouldn't worry us. Sagi has convincingly argued that it is incorrect to deny the logical nature of unicorn negation on wider modal or epistemic grounds. We would only be worried if unicorn negation turns out to be non-topic-neutral. But if we understand topic-neutrality in the sense of Discrimination—as fits most naturally with isomorphism-invariance—then unicorn negation appears to be topic-neutral. Further, we have noted that intuitions about such cases are far from clear and that the major argument for the non-logical nature of unicorn negation—it implies 'there are no unicorns'—fails. The problem of predicates such as 'is a unicorn' remains. Here we pointed to two approaches, the first based on formalization, of which we were sceptical, the second and more promising one based on Sher's idea that the meanings of the logical constants are exhausted by their denotation.

We've now considered four different sorts of objection to isomorphism invariance as an approach to logical extensions and logical constants. In order, these were: the heterogeneity objection (Chapter 8); the overgeneration objection (Chapter 9); the absoluteness objection (Chapter 10); and the intensional objection (this chapter). In response to the heterogeneity objection, we argued that closure under definability entails a large degree of heterogeneity, and that neither Feferman nor Bonnay's alternative invariantism is as well-motivated as ours. In response to overgeneration, we argued that our account does not overgenerate; it does not problematically assimilate logic to mathematics. In response to absoluteness, we showed that this objection relied on a dubious metaphysics of sets. And we've just responded to the intensional objection, which in any case doesn't affect our argument for the LooGooS Hypothesis.

Conclusion

We have defended the L∞G∞S Hypothesis that the one true logic is of sort ∞∞. Our argument has been as follows.

First, we are concerned with logical consequence as it is used in natural language, with pride of place given to its more technical domains such as mathematics, science, and the law. In Chapter 1, we found that the intuitive concept of logical consequence has many features and that they are all controversial. The most widely accepted intuitive feature is truth-preservation. Beyond that, the most-discussed features are formality and necessity, although these are far from settled.

One reaction to this situation is logical pluralism, which we first discussed in Chapter 2. On the most plausible versions of this view, such as Beall and Restall's and Shapiro's, there are many correct logics, which correspond to different precisifications of the intuitive notion. Unfortunately, as we argued in Chapter 3, logical pluralism is unstable.

Logical pluralists, like everyone, will sometimes want to reason metatheoretically, such as when they are trying to convince others of the truth of pluralism. In the metatheory, however, they must reason in ways acceptable to all of the logics they countenance. If, like Shapiro, they want to endorse any logic with an interesting application, then logical nihilism threatens: there is little, if anything, accepted by all logicians. The pluralist could, like Beall and Restall, restrict the range of countenanced logics to, for example, those with certain features settled by the tradition. This is where the discussion of Chapter 1 finds its bite: *nothing* is settled by the tradition. To think otherwise is to introduce an arbitrary stopping point. We are left with a spectrum of logics countenanced. At one end is monism, at the other Shapiro's eclectic pluralism, with Beall and Restall somewhere in between. Eclectic pluralism is metatheoretically incoherent. Intermediate positions lack motivation. We are left with monism.

Of course, monists cannot hope to capture all of logic's intuitive features, since some are in tension with each other. Instead, the monist should take some strand from the tradition and attempt its explication, putting up the account for evaluation by the usual abductive criteria. Like Tarski, we took the formality of logic and ran with it. In Chapter 6, we tidied this up using MacFarlane's notion of 2-formality into an intuitive understanding of formality as topic-neutrality. On this view, consequence is formal in the sense of schematic. This picture is incomplete without an account of the logical constants—for these are what is held fixed when determining forms—or their extensions.

One True Logic: A Monist Manifesto. Owen Griffiths and A.C. Paseau, Oxford University Press.
© Owen Griffiths and A.C. Paseau 2022. DOI: 10.1093/oso/9780198829713.003.0012

In Part II, we filled part of this gap in the account with a story about logical relations (i.e. extensions). All relations that are topic-neutral are logical, and the best way to spell out topic-neutrality is by means of isomorphism invariance. And for any logical relation, there must be a logical constant denoting it. This leads to a logic of sort $\infty\infty$ and thus to the L∞G∞S Hypothesis, articulated in Chapter 4.

Isomorphism invariance is one way to argue for the L∞G∞S Hypothesis, as we saw in Chapter 6, but not the only one. A distinct argument was presented in Chapter 5. There, we considered the central motivation for logic: to underwrite the validity or invalidity of arguments. We gave our argument $\mathcal{A}$ for the conclusion that there are infinitely many planets and saw that first-order logic cannot accommodate its validity. Second-order logic can, but there are generalizations that are missed here. This gave us a general recipe for cooking up valid arguments that will be missed by any logic weaker than one of sort $\infty\infty$. We also knocked down Quine's concurrence-of-ideas argument for first-order logic.

At this point, the L∞G∞S Hypothesis looked attractive. It had a powerful 'top-down' argument stemming from logic's formal character and a series of compelling 'bottom-up' arguments. We rounded off our discusion by raising and defeating some objections in Part III.

In Chapter 8, we cast a critical eye over some rival forms of invariantism and thereby saw off the heterogeneity objection. In Chapter 9, we saw that isomorphism invariance does not overgenerate. It has been charged with finding mathematical claims such as CH to be logically true, if true; or logically false, if false. We gave several precisifications of this argument and found them all wanting. In Chapter 10, we saw that the absoluteness objection has no bite, since it lacks plausible motivation. In Chapter 11, we considered intensional objections and especially that based on unicorn negation, $\mathcal{U}$. We fended off these objections, even though not required for the defence of our thesis in Part II.

Overall, then, we have argued for the truth of the L∞G∞S Hypothesis. It leads to a radically new approach to logic. While not singling out a particular logic, it has shown us that logic must be strongly infinitary (i.e. have maximally infinitary sort). The task now is to narrow down the other dimensions of the one true logic.

Bibliography

A. Ahmed (2012), 'Hale on the necessity of necessity', *Mind* 109, pp. 81–92.

Aristotle, *Prior Analytics*, 1938 ed., Loeb Classical Library: Harvard University Press.

J.T. Baldwin (2018), *Model Theory and the Philosophy of Mathematical Practice*, Cambridge University Press.

T.W. Barrett and H. Halvorson (2017), 'Quine's Conjecture on Many-Sorted Logic', *Synthese* 194, pp. 3563–82.

J. Barwise (1977), 'An Introduction to First-Order Logic', in J. Barwise (ed.), *Handbook of Mathematical Logic*, Elsevier, pp. 5–46.

J. Barwise and R. Cooper (1981), 'Generalized Quantifiers and Natural Language', *Linguistics and Philosophy* 4, pp. 159–219.

J. Barwise and S. Feferman (1985), *Model-Theoretic Logics*, Springer-Verlag.

J. Barwise and J. Perry (1983), *Situations and Attitudes*, MIT Press.

D. Batens (1985), 'Meaning, Acceptance and Dialectics', in J.C. Pitt (ed.), *Change and Progess in Modern Science*, Reidel.

Jc Beall (2009), *Spandrels of Truth*, Oxford University Press.

Jc Beall and G. Restall (2000), 'Logical Pluralism', *Australasian Journal of Philosophy* 78, pp. 475–93.

Jc Beall and G. Restall (2001), 'Defending Logical Pluralism', in J. Woods and B. Brown (eds), *Logical Consequence: Rival Approaches*, Hermes Science, pp. 1–22.

Jc Beall and G. Restall (2006), *Logical Pluralism*, Oxford University Press.

Jc Beall, M. Glanzberg, and D. Ripley (2018), *Formal Theories of Truth*, Oxford University Press.

J.L. Bell (2008), *A Primer of Infinitesimal Analysis*, Cambridge University Press.

J.L. Bell (2016), 'Infinitary Logic', *The Stanford Encyclopedia of Philosophy*, E. N. Zalta (ed.).

P. Blanchette (2000), 'Models and Modality', *Synthese* 124, pp. 45–72.

P. Blanchette (2001), 'Logical Consequence', in L. Goble (ed.), *The Blackwell Guide to Philosophical Logic*, pp. 115–35.

I. Boh (2001), 'Consequence in the Post-Ockham Period', in M. Yrjönsuuri (ed.), *Medieval Formal Logic*, Kluwer Press, pp. 147–82.

B. Bolzano (1837), *Wissenschaftslehre*, Blackwell Press (1972). References to 1972 partial translation by R. George, *Theory of Science*.

D. Bonnay (2008), 'Logicality and Invariance', *The Bulletin of Symbolic Logic* 14, pp. 29–68.

G. Boolos (1975), 'On Second-Order Logic', repr. in his *Logic, Logic, and Logic* (1998), Harvard University Press, pp. 37–53.

G. Boolos (1984), 'To Be is to Be a Value of Variable (or to Be Some Values of Some Variables)', repr. in his *Logic, Logic, and Logic* (1998), Harvard University Press, pp. 54–72.

J. Buridan, *The Treatise on Consequences*, Reidel Press. References to 1985 translation by P. King.

T. Button and S. Walsh (2018), *The Philosophy of Model Theory*, Oxford University Press.

J. Cargile (2010), 'Logical Form', in A. Oliver and J. Lear (eds), *The Force of Argument: Essays in Honour of Timothy Smiley*, Routledge, pp. 48–67.

R. Carnap (1934), *Logische Syntax der Sprache*, Littlefield Adams Press. References to 1959 translation by A. Smeaton.

R. Carnap (1947), *Meaning and Necessity* (2nd ed.), University of Chicago Press.

C.C. Chang and H. Jerome Keisler (1990), *Model Theory* (3rd ed.), North-Holland.

N. Chomsky (1957), *Syntactic Structures*, Mouton and Co.

R. Cook (2010), 'Let a Thousand Flowers Bloom: A Tour of Logical Pluralism', *Philosophy Compass* 5/6, pp. 492–504.

A. Cotnoir (2018), 'Logical Nihilism', in N. Kellen, N.J.L.L. Pedersen, and J. Wyatt (eds), *Pluralisms in Truth and Logic*, Palgrave Macmillan, pp. 301–29.

D. Davidson (1965), 'Theories of Meaning and Learnable Languages', refs to reprint in *Inquiries into Truth and Interpretation*, Clarendon Press, pp. 3–16.

D. Davidson (1967), 'The logical form of action sentences', refs to reprint in *Essays on Actions and Events*, Clarendon Press, pp. 105–22.

M. Detlefsen (2008), 'Purity as an Ideal of Proof', in P. Mancosu (ed.), *The Philosophy of Mathematical Practice*, Oxford University Press, pp. 179–97.

M.A. Dickmann (1985), 'Larger Infinitary Languages', in Barwise and Feferman (1985), pp. 317–63.

V.H. Dudman (1988), 'Indicative and Subjunctive', *Analysis* 48, pp. 113–22.

C. Dutilh Novaes (2005), 'In Search of the Intuitive Notion of Logical Consequence', in *LOGICA Yearbook 2004*, Filosofia Press, pp. 109–23.

C. Dutilh Novaes (2011), 'The Different Ways in which Logic is (Said to be) Formal', *History and Philosophy of Logic* 32, pp. 303–32.

C. Dutilh Novaes (2012), *Formal Languages in Logic: A Philosophical and Cognitive Analysis*, Cambridge University Press.

C. Dutilh Novaes (2014), 'The Undergeneration of Permutation Invariance as a Criterion for Logicality', *Erkenntnis* 79, pp. 81–97.

C. Dutilh Novaes (2021), *The Dialogical Roots of Deduction*, Cambridge University Press.

C. Dutilh Novaes and S. Uckelman (2016), 'Obligationes', in C. Dutilh Novaes and S. Read (eds), *The Cambridge Companion to Medieval Logic*, Cambridge University Press, pp. 370–95.

J. Earman (1995), *Bangs, Crunches, Whimpers, and Shrieks: Singularities and Acausalities in Relativistic Spacetimes*, Oxford University Press.

H.-D. Ebbinghaus, J. Flum, and W. Thomas (1994), *Mathematical Logic* (2nd ed.), Springer.

M. Eklund (1996), 'On How Logic Became First-Order', *Nordic Journal of Philosophical Logic* 1, pp. 147–167.

A. Einstein (1921), 'Geometrie und Erfahrung', transl. by S. Bargmann as 'Geometry and Experience' and repr. in his *Ideas and Opinions* (1954), Condor, pp. 232–46.

L. Estrada-González (2011), 'On the Meaning of Connectives (A Propos of a Non-Necessitarian Challenge)', *Logica Universalis* 5, pp. 115–26.

J. Etchemendy (1983), 'The Doctrine of Logic as Form', *Linguistics and Philosophy* 6, pp. 319–34.

J. Etchemendy (1990), *The Concept of Logical Consequence*, Harvard University Press.

J. Etchemendy (2008), 'Reflections on consequence', in D. Patterson (ed.), *New Essays on Tarski and Philosophy*, Oxford University Press, pp. 263–299.

W. Ewald (ed.) (1996), *From Kant to Hilbert: Readings in the Foundations of Mathematics*, 2 volumes, Clarendon Press.

S. Feferman (1999), 'Logic, Logics and Logicism', *Notre Dame Journal of Formal Logic* 40, pp. 31–54.

S. Feferman (2000), 'Why the programs for new axioms need to be questioned', *Bulletin of Symbolic Logic* 6, pp. 401–13.

S. Feferman (2010a), 'Set-theoretical Invariance Criteria for Logicality', *Notre Dame Journal of Formal Logic* 51, pp. 3–20.

S. Feferman (2010b), 'On the Strength of Some Semi-constructive Set Theories', in *Proofs, Categories and Computations: Essays in Honor of Grigori Mints*, S. Feferman and W. Sieg (eds), College Publications, pp. 109–29.

S. Feferman (2011), 'Is the Continuum Hypothesis a Definite Mathematical Problem?', unpublished ms.

H. Field (1980), *Science Without Numbers: A Defense of Nominalism*, Princeton University Press.

H. Field (1989), *Realism, Mathematics and Modality*, Oxford Blackwell Press.

H. Field (2008), *Saving Truth from Paradox*, Oxford University Press.

S. Florio and L. Incurvati (2019), 'Metalogic and the Overgeneration Argument', *Mind* 128, pp. 761–93.

G. Frege (1885), 'Uber formale Theorien der Arithmetik' transl. by E.-H.W. Kluge as 'On Formal Theories of Arithmetic' and repr. in B. McGuinness (ed.), *Collected Papers on Mathematics, Logic and Philosophy* (1984), Blackwell Press, pp. 112–21.

G. Frege (1918), 'Der Gedanke. Eine logische Unterschung', transl. by P. Geach and R.H. Stoothoff as 'Thoughts' and repr. in B. McGuinness (ed.), *Collected Papers on Mathematics, Logic and Philosophy* (1984), Blackwell Press, pp. 351–72.

M. Friedman (1974), 'Explanation and Scientific Understanding', *The Journal of Philosophy* 71, pp. 5–19.

P. Godfrey-Smith (2003), *Theory and Reality*, University of Chicago Press.

R. Goldblatt (1984), *Topoi: The Categorial Analysis of Logic* (2nd ed.), North-Holland.

M. Gómez-Torrente (1996), 'Tarski on logical consequence', *Notre Dame Journal of Formal Logic* 37, pp. 125–51.

M. Gómez-Torrente (1998), 'Logical Truth and Tarskian Logical Truth', *Synthese* 117, pp. 375–408.

M. Gómez-Torrente (2002), 'The Problem of Logical Constants', *The Bulletin of Symbolic Logic* 8, pp. 1–37.

M. Gómez-Torrente (2021), 'The Problem of Logical Constants and the Semantic Tradition: From Invariantist Views to a Pragmatic Account', in Sagi and Woods (2021), pp. 35–54.

J.B. Gould (1970), *The Philosophy of Chrysippus*, Vol. XVII of *Philosophia Antiqua*, Leiden Press.

O. Griffiths (2013), 'Problems for Logical Pluralism', *History and Philosophy of Logic* 34, pp. 170–82.

O. Griffiths (2014), 'Formal and Informal Consequence', *Thought* 3, pp. 9–20.

O. Griffiths and A. Ahmed (2021), 'Introducing Identity', *Journal of Philosophical Logic* 50, pp. 1449–69.

O. Griffiths and A.C. Paseau (2016), 'Isomorphism Invariance and Overgeneration', *Bulletin of Symbolic Logic* 22, pp. 482–503.

V. Halbach (2010), *The Logic Manual*, Oxford University Press.

V. Halbach (2014), *Axiomatic Theories of Truth*, Cambridge University Press.

J.D. Hamkins (2012), 'The Set-theoretic Multiverse', *The Review of Symbolic Logic* 5, pp. 416–49.

J.D. Hamkins (2015), 'Is the Dream Solution of the Continuum Hypothesis Attainable?', *Notre Dame Journal of Formal Logic* 56, pp. 135–45.

W.H. Hanson (1997), 'The Concept of Logical Consequence', *The Philosophical Review* 106, pp. 365–409.

G. Harman (1986), *Change in View*, MIT Press.

G. Harman (2009), 'Field on the Normative Role of Logic', *Proceedings of the Aristotelian Society* 109, pp. 333–5.

M.D. Hauser, N. Chomsky, and W.T. Fitch (2002), 'The Faculty of Language: What Is It, Who Has It, and How Did It Evolve?', *Science* 298, pp. 1569–79.

G. Hellman (1989), *Mathematics Without Numbers*, Oxford University Press.

G. Hellman (2006), 'Mathematical Pluralism: The Case of Smooth Infinitesimal Analysis', *Journal of Philosophical Logic* 35, pp. 621–51.

C.G. Hempel and P. Oppenheim (1948), 'Studies in the Logic of Explanation', *Philosophy of Science* 15, pp. 135–75.

D. Hilbert (1905), 'Über die Grundlagen der Logik und der Arithmetik', transl. by B. Woodward as 'On the Foundations of Logic and Arithmetic' and repr. in van Heijenoort (1967), pp. 129–38.

O. Hjortland (2013), 'Logical Pluralism, Meaning-Variance, and Verbal Disputes', *Australasian Journal of Philosophy* 91, pp. 355–73.

A. Iacona (2010), 'Truth Preservation in any Context', *American Philosophical Quarterly* 47, pp. 191–9.

B. Jackson (2006), 'Logical Form: Classical Conception and Recent Challenges', *Philosophy Compass* 1, pp. 303–16.

T. Jech (1973), *The Axiom of Choice*, North-Holland.

I. Kant (1800), *Jäsche Logic (Logic: A Manual for Lectures)*, in Kant Ak: IX, references to *Lectures on Logic* (1992), transl. by M. Young, Cambridge University Press.

D. Kaplan (1989), 'Demonstratives: an Essay on the Semantics, Logic, Metaphysics and Epistemology of Demonstratives', in J. Almog, J. Perry, and H. Wettstein (eds), *Themes from Kaplan* (1989), Oxford University Press, pp. 481–563.

M. Kaufman (1985), 'The quantifier "there exist uncountably many" and some of its relatives', in J. Barwise and S. Feferman (eds), *Model-Theoretic Logics* (1985), Springer-Verlag, pp. 123–76.

E.L. Keenan and L.S. Moss (2016), *Mathematical Structures in Language*, CSLI.

H.J. Keisler (1970), 'Logic with the quantifier "there exist uncountably many"', *Annals of Mathematical Logic* 1, pp. 1–93.

H.J. Keisler (1971), *Model Theory for Infinitary Logic*, North-Holland.

P. King (2001), 'Consequence as Inference: Mediaeval Proof Theory 1300–1350', in M. Yrjönsuuri (ed.), *Medieval Formal Logic*, Kluwer Press, pp. 117–46.

P. Kitcher (1981), 'Explanatory Unification', *Philosophy of Science* 48, pp. 507–31.

P. Kitcher (1989), 'Explanatory Unification and the Causal Structure of the World', in P. Kitcher and W. Salmon (eds), *Scientific Explanation*, University of Minnesota Press, pp. 410–505.

W. Kneale (1961), 'Universality and necessity', *The British Journal for the Philosophy of Science* 46, pp. 89–102.

W. Kneale and M. Kneale (1962), *The Development of Logic*, Clarendon Press.

G. Kreisel (1967), 'Informal Rigour and Completeness Proofs', in I. Lakatos (ed.), *Problems in the Philosophy of Mathematics*, North-Holland, pp. 138–71.

S. Kripke (1980), *Naming and Necessity*, Harvard University Press.

K. Kunen (2013), *Set Theory*, College Publications.

M. Lange (2017), *Because Without Cause*, Oxford University Press.

D.T. Langendoen and P.M. Postal (1984), *The Vastness of Natural Languages*, Basil Blackwell.

A. Lindenbaum and A. Tarski (1935), 'Über die Beschränktheit der Ausdrucksmittel deduktiver Theorien', transl. as 'On the limitations of the means of expression of deductive theories' by J.H. Woodger and repr. in *Logic, Semantics, MetaMathematics*, 1983, J. Corcoran (ed.), Hackett, pp. 384–92.

P. Lindström (1966), 'First order predicate logic with generalized quantifiers', *Theoria* 35, pp. 186–95.

P. Lindström (1969), 'On Extensions of Elementary Logic', *Theoria* 35, pp. 1–11.

Ø. Linnebo (2013), 'The potential hierarchy of sets', *The Review of Symbolic Logic* 6, pp. 205–28.

L. Löwenheim (1915), 'Über Möglichkeiten im Relativkalkül', transl. as 'On possibilities in the calculus of relations' by S. Bauer-Mengelberg and repr. in van Heijenoort (1967), pp. 228–51.

T. McCarthy (1981), 'The Idea of a Logical Constant', *Journal of Philosophy* 78, pp. 499–523.

T. McCarthy (1987), 'Modality, invariance and logical truth', *Journal of Philosophical Logic* 16, pp. 423–43.

J. MacFarlane (2000), *What Does it Mean to Say that Logic is Formal?*, PhD thesis, University of Pittsburgh.

J. MacFarlane (2015), 'Logical Constants', in E. Zalta (ed.), *The Stanford Encyclopedia of Philosophy*.

V. McGee (1985), 'A Counterexample to Modus Ponens', *The Journal of Philosophy* 82, pp. 462–71.

V. McGee (1992), 'Two Problems for Tarski's Theory of Consequence', *Proceedings of the Aristotelian Society* 92, pp. 273–92.

V. McGee (1996), 'Logical Operations', *Journal of Philosophical Logic* 25, pp. 567–80.

M. Magidor and J. Väänänen (2011), 'On Löwenheim-Skolem-Tarski Numbers for Extensions of First-Order Logic', *Journal of Mathematical Logic* 11, pp. 87–113.

P. Mancosu (2010), 'Fixed- versus Variable-domain Interpretations of Tarski's Account of Logical Consequence', *Philosophy Compass* 5, pp. 745–59.

B. Mates (1953), *Stoic Logic*, University of California Berkeley Press. References to 1961 reprint.

B. Mates (1972), *Elementary Logic* (2nd ed.), New York Press.

F.I. Mautner (1946), 'An Extension of Klein's Erlanger Program: Logic as Invariant-Theory', *American Journal of Mathematics* 68, pp. 345–84.

R.K. Meyer and K. Lambert (1968), 'Universally Free Logic and Standard Quantification Theory', *The Journal of Symbolic Logic* 33, pp. 8–26.

P. Milne (2007), 'Existence, Freedom, Identity, and the Logic of Abstractionist Realism', *Mind* 116, pp. 23–53.

R. Montague (1974), *Formal Philosophy*, R. Thomason (ed.), Yale University Press.

G.H. Moore (1980), 'Beyond First-order Logic: the Historical Interplay between Mathematical Logic and Axiomatic Set Theory', *History and Philosophy of Logic* 1, pp. 95–127.

G.H. Moore (1988), 'The Emergence of First-order Logic', in W. Aspray and P. Kitcher (eds), *History and Philosophy of Modern Mathematics: Minnesota Studies in the Philosophy of Science vol XI*, University of Minnesota Press, pp. 95–135.

G.H. Moore (1990), 'Proof and the Infinite', *Interchange* 21, pp. 46–60.

G.H. Moore (1997), 'The Prehistory of Infinitary Logic: 1885–1955', in M.L. Dalla Chiara, K. Doets, D. Mundici, and J. van Benthem (eds), *Structure and Norms in Science: Volume Two of the Tenth International Congress of Logic, Methodology and Philosophy of Science, Florence, August 1995* (*Synthese* Library vol. 260), Kluwer, pp. 105–23.

A. Mostowski (1957), 'On a Generalisation of Quantifiers', *Fundamenta Mathematicae* 44, pp. 12–36.

E. Müller (1910), *Abriss der Algebra der Logik* vol. 2, Teubner.

C. Normore (1993), 'The Necessity in Deduction: Cartesian Inference and its Medieval Background', *Synthese* 96, pp. 437–54.

A.D. Oliver (1999), 'A Few More Remarks on Logical Form', *Proceedings of the Aristotelian Society* 99, pp. 247–72.

A.D. Oliver and T.J. Smiley (2013), *Plural Logic*, Oxford University Press.

A.C. Paseau (2007), Review of *Logical Pluralism* by JC Beall and G. Restall, *Mind* 116, pp. 391–6.

A.C. Paseau (2010), 'Pure Second-Order Logic with Second-Order Identity', *Notre Dame Journal of Formal Logic* 51, pp. 351–60.

A.C. Paseau (2014), 'The Overgeneration Argument(s): A Succinct Refutation', *Analysis* 74, pp. 40–7.

A.C. Paseau (2019a), 'A Measure of Inferential-Role Preservation', *Synthese* 196 (2019), pp. 2621–42.

A.C. Paseau (2019b), 'Capturing Consequence', *The Review of Symbolic Logic* 12, pp. 271–95.

A.C. Paseau (2020), 'Non-metric propositional similarity', *Erkenntnis*, vol. and pp. tbc.

A.C. Paseau (2021a), '*Logos*, Logic and Maximal Infinity', *Religious Studies*, vol. and pp. tbc.

A.C. Paseau (2021b), 'Propositionalism', *The Journal of Philosophy* 118, pp. 430–49.

A.C. Paseau (forthcoming), *What is Mathematics About?*, Oxford University Press.

A.C. Paseau and O. Griffiths (2021a), 'Propositional logics of truth by logical form', in Sagi and Woods (2021), pp. 160–85.

A.C. Paseau and O. Griffiths (2021b), 'Is English Consequence Compact?', *Thought* 10, pp. 160–85.

G. Payette and N. Wyatt (2018), 'How do logics explain?', *Australasian Journal of Philosophy* 96, pp. 157–67.

S. Peters and D. Westerståhl (2006), *Quantifiers in Language and Logic*, Oxford University Press.

D. Pinheiro Fernandes (2018), 'Translations: generalizing relative expressiveness between logics', arXIv:1706.08481v1.

K. Popper (1948), 'On the Theory of Deduction', *Indagationes Mathematicae* 10, pp. 44–54 (part I) & pp. 111–20 (part II). [*Koninklije Nederlandse Akademie van Wetenschappen, Proceedings of the section of sciences* 51, pp. 173–83 (part I) & pp. 322–31 (part II).]

E.L. Post (1921), 'Introduction to a general theory of elementary propositions', repr. in van Heijenoort (1967), pp. 265–83.

G. Priest (1995), 'Etchemendy and Logical Consequence', *Canadian Journal of Philosophy* 25, pp. 283–92.

G. Priest (2001), 'Logic: One or Many?', in J. Woods and B. Brown (eds), *Logical Consequences: Rival Approaches*, Hermes Scientific Publishers, pp. 23–38.

G. Priest (2006), *Doubt Truth to be a Liar*, Oxford University Press.

W.V. Quine (1951), 'Two Dogmas of Empiricism', *The Philosophical Review* 60, pp. 20–43.

W.V. Quine (1952), *Methods of Logic* Harvard University Press.

W.V. Quine (1953), 'Three Grades of Modal Involvement', repr. in his *The Ways of Paradox* 1966, Harvard University Press, pp. 156–74.

W.V. Quine (1954), 'Quantification and the Empty Domain', *The Journal of Symbolic Logic* 19, pp. 177–9.

W.V. Quine (1970), *Philosophy of Logic*, Harvard University Press.

W.V. Quine (1980), 'Grammar, Truth and Logic' in S. Kanger and S. Öhman (eds), *Philosophy and Grammar*, Reidel Press, pp. 17–28.

W.V. Quine (1982), *Methods of Logic* (4th ed.), Harvard University Press.

F.P. Ramsey (1925), 'The Foundations of Mathematics', *Proceedings of the London Mathematical Society* (Series 2) 25, pp. 338–84.

S. Read (1994), 'Formal and Material Consequence', *Journal of Philosophical Logic* 23, pp. 247–65.

S. Read (2004), 'Identity and Harmony', *Analysis* 64, pp. 113–19.

S. Read (2012), 'The Medieval Theory of Consequence', *Synthese* 187, pp. 899–912.

S. Read (2016), 'Harmonic Inferentialism and the Logic of Identity', *The Review of Symbolic Logic* 9, pp. 408–20.

M. Resnik (1988), 'Second-order logic still wild', *The Journal of Philosophy* 85, pp. 75–87.

G. Restall (2000), *An Introduction to Substructural Logics*, Routledge.

G. Restall (2002), 'Carnap's Tolerance, Meaning and Logical Pluralism', *The Journal of Philosophy* 99, pp. 436–43.

I. Rumfitt (2015), *The Boundary Stones of Thought*, Oxford University Press.

B. Russell (1905), 'Necessity and Possibility', reprinted in A. Urquhart (ed.), *The Collected Papers of Bertrand Russell, Vol. 4* (1994), Routledge, pp. 508–20.

B. Russell (1908), 'Mathematical Knowledge as Based on the Theory of Types', *American Journal of Mathematics* 30, pp. 222–62.

B. Russell (1914), *Our Knowledge of the External World as a Field for Scientific Method in Philosophy*, Open Court. References to the 1993 reprinting by Routledge.

B. Russell (1919), *Introduction to Mathematical Philosophy*, Routledge Press.

B. Russell (1936), 'The Limits of Empiricism', *Proceedings of the Aristotelian Society* 36, pp. 131–50.

G. Russell (2008), 'One true logic?', *Journal of Philosophical Logic* 37, pp. 593–611.

G. Russell (2014), 'Logical Pluralism' in E. Zalta (ed.), *The Stanford Encyclopedia of Philosophy*.

G. Russell (2018), 'Logical Nihilism: Could There be No Logic?', *Philosophical Issues*, 28, pp. 308–24.

G. Sagi (2015), 'The Modal and Epistemic Arguments Against the Invariance Criterion for Logical Terms', *The Journal of Philosophy* 112, pp. 159–67.

G. Sagi and J. Woods (eds) (2021), *The Semantic Conception of Logic: Essays on Consequence, Invariance, and Meaning*, Cambridge University Press.

M. Sainsbury (2001), *Logical Forms* (2nd ed.), Oxford Blackwell Press.

W. Salmon (1998), *Causality and Explanation*, Oxford University Press.

C. Scambler (2020), 'An Indeterminate Universe of Sets', *Synthese* 197, pp. 545–73.

A. Sereni and M. P. Sforza Fogliani (2020), 'How to Water a Thousand Flowers: On the Logic of Logical Pluralism', *Inquiry* 63, pp. 347–70.

S. Shapiro (1991), *Foundations without Foundationalism*, Clarendon Press.

S. Shapiro (1998), 'Logical Consequence: Models and Modality', in M. Schirn (ed.), *The Philosophy of Mathematics Today*, Oxford University Press, pp. 131–56.

S. Shapiro (2005), 'Logical consequence, proof theory, and model theory', in S. Shapiro (ed.), *The Oxford Handbook of Philosophy of Mathematics and Logic*, Oxford University Press, pp. 651–70.

S. Shapiro (2014), *Varieties of Logic*, Oxford University Press.

G. Sher (1991), *The Bounds of Logic*, MIT Press.

G. Sher (1996), 'Did Tarski Commit Tarski's Fallacy?', *The Journal of Symbolic Logic* 61, pp. 653–86.

G. Sher (2001), 'Is Logic a Theory of the Obvious?', in J. Woods and B. Brown (eds), *Logical Consequences: Rival Approaches*, Hermes Scientific Publishers, pp. 55–79.

G. Sher (2008), 'Tarski's thesis', in D. Patterson (ed.), *New Essays on Tarski and Philosophy*, Oxford University Press, pp. 300–39.

G. Sher (2013), 'The Foundational Problem of Logic', *Bulletin of Symbolic Logic* 19, pp. 145–98.

G. Sher (2016), *Epistemic Friction: An Essay on Knowledge, Truth, and Logic*, Oxford University Press.

G. Sher (2021), 'Invariance and Logicality in Perspective', in Sagi and Woods (2021), pp. 10–25.

D.J. Shoesmith and T.J. Smiley (1978), *Multiple-conclusion Logic*, Cambridge University Press.

S. Simpson (2010), *Subsystems of Second-order Logic* (2nd ed.), ASL.

T. Skolem (1920), 'Logisch-kombinatorische Untersuchungen über die Erfüllbarkeit oder Beweisbarkeit mathematischer Sätze nebst einem Theoreme über dichte Mengen', transl. by S. Bauer-Mengelberg as 'Logico–Combinatorial Investigations in the Satisfiability or Provability of Mathematical Propositions: A Simplified Proof of a Theorem by L. Löwenheim and Generalizations of the Theorem' and repr. in van Heijenoort (1967), pp. 252–63.

T. Skolem (1922), 'Einige Bemerkungen zur axiomatischen Begründung der Mengenlehre', transl. by S. Bauer-Mengelberg as 'Some Remarks on Axiomatized Set Theory' and repr. in van Heijenoort (1967), pp. 290–301.

T. Skolem (1923), 'Begründung der elementaren Arithmetik durch die rekurrierende Denkweise ohne Anwendung scheinbarer Veränderlichen mit unendlichem Ausdehnungsbereich', transl. by S. Bauer-Mengelberg as 'The Foundations of Elementary Arithmetic Established by Means of the Recursive Mode of Thought, Without the Use of Apparent Variables Ranging over Infinite Domains' and repr. in van Heijenoort (1967), pp. 302–333.

T.J. Smiley (1998), 'Conceptions of Consequence', in E.J. Craig (ed.), *Routledge Encyclopedia of Philosophy*, Routledge.

P. Smith (2011), 'Squeezing arguments', *Analysis* 71, pp. 22–30.

S. Soames (1999), Appendix A (pp. 117–36) to Chapter 4 of *Understanding Truth*, Oxford University Press.

E. Stei (2020), 'Disagreement about logic from a pluralist perspective', *Philosophical Studies* 177, pp. 3329–50.

F. Steinberger (2019), 'Logical Pluralism and Logical Normativity', *Philosophers' Imprint* 19, pp. 1–19.

P. Steinkrüger (2015), 'Aristotle's assertoric syllogistic and modern relevance logic', *Synthese*, pp. 1413–44.

P. Strawson (1952), *Introduction to Logical Theory*, Methuen.

A. Tarski (1935), 'Der Wahrheitsbegriff in den formalisierten Sprachen', transl. from the Polish by L. Blaustein and transl. from the German by J. Woodger as "The concept of truth in formalized languages' and repr. in J. Corcoran (ed.), Tarski's *Logic, Semantics, Metamathematics* (1983, 2nd ed.), Clarendon Press, pp. 152–278.

A. Tarski (1936), 'Über den Begriff der logischen Folgerung', transl. by J. Woodger as 'On the Concept of Logical Consequence' and repr. in J. Corcoran (ed.), Tarski's *Logic, Semantics, Metamathematics*, 2nd ed. (1983), Clarendon Press, pp. 409–20.

A. Tarski (1937), *Introduction to Logic and the Methodology of Deductive Sciences*, Dover. References to 1995 reprint.

A. Tarski (1986), 'What are Logical Notions?', J. Corcoran (ed.), *History and Philosophy of Logic* 7, pp. 143–54.

A. Tarski (2002), 'On the concept of following logically', *History and Philosophy of Logic* 23, pp. 155–96. Translated from the Polish by M. Stroińska and D. Hitchcock.

A. Tarski and R. Vaught (1956), 'Arithmetical extensions of relational systems', *Compositio Mathematica* 13, pp. 81–102.

L. Tharp (1975), 'Which logic is the right logic?', *Synthese* 31, pp. 1–31.

A. S. Troelstra and D. van Dalen (1988), *Constructivism in Mathematics* (2 vols), North-Holland.

R. Trueman (2012), 'Neutralism Within the Semantic Tradition', *Thought* 3, pp. 246–51.

J. Väänänen (2004), 'Barwise: Abstract Model Theory and Generalized Quantifiers', *Bulletin of Symbolic Logic* 10, pp. 37–53.

J. Väänänen (2012), 'Second order logic or set theory', *Bulletin of Symbolic Logic* 18, pp. 91–121.

J. van Benthem (1989), 'Logical constants across varying types', *Notre Dame Journal of Formal Logic* 30, pp. 315–42.

B. van Fraassen (1980), *The Scientific Image*, Oxford University Press.

J. van Heijenoort (ed.) (1967), *From Frege to Gödel: A Source Book in Mathematical Logic, 1879–1931*, Harvard University Press.

R. Vaught (1964), 'The completeness of logic with the added quantifier "there are uncountably many"', *Fundamenta Mathematicae* 54, pp. 303–4.

K. Warmbröd (1999), 'Logical Constants', *Mind* 108, pp. 503–38.

J.O. Weatherall (2016), 'Understanding Gauge', *Philosophy of Science* 83, pp. 1039–49.

T. Williamson (2007), *The Philosophy of Philosophy*, Oxford University Press.

T. Williamson (2011), 'Reply to Boghossian', *Philosophy and Phenomenological Research* 82, pp. 498–506.

T. Williamson (2013), *Modal Logic as Metaphysics*, Oxford University Press.

T. Williamson (2014), 'Logic, Metalogic and Neutrality', *Erkenntnis* 79, pp. 211–31.

T. Williamson (2017), 'Semantic Paradoxes and Abductive Methodology', in B. Armour-Garb (ed.), *The Relevance of the Liar*, Oxford University Press, pp. 325–46.

L. Wittgenstein (1921), *Tractatus Logico-Philosophicus*, various eds.

J. Woods (2019), 'Logical Partisanhood', *Philosophical Studies* 176, pp. 1203–24.

J. Woodward (2003), *Making Things Happen: A Theory of Causal Explanation*, Oxford University Press.

C. Wright (2008), 'Relativism about truth itself: haphazard thoughts about the very idea', in M. García-Carpintero and M. Kölbel (eds), *Relative Truth*, pp. 157–86.

B.-U. Yi (2006), 'The Logic and Meaning of Plurals. Part II', *Journal of Philosophical Logic* 35, pp. 239–88.

E. Zermelo (1930), 'Über Grenzzahlen und Mengenbereiche: Neue Untersuchungen über die Grundlagen die Mengenlehre', trans. by M. Hallett as 'On Boundary Numbers and Domains of Sets: New Investigations in the Foundations of Set Theory' and repr. in Ewald (1996), pp. 1208–33.

E. Zermelo (1932), 'Über mathematische Systeme und die Logik des Unendlichen', *Forschungen und Fortschritte* 8, pp. 6–7.

P. Ziff (1974), 'The Number of English Sentences', *Foundations of Language* 11, pp. 519–32.

Index